Springer Theses

Recognizing Outstanding Ph.D. Research

Aims and Scope

The series "Springer Theses" brings together a selection of the very best Ph.D. theses from around the world and across the physical sciences. Nominated and endorsed by two recognized specialists, each published volume has been selected for its scientific excellence and the high impact of its contents for the pertinent field of research. For greater accessibility to non-specialists, the published versions include an extended introduction, as well as a foreword by the student's supervisor explaining the special relevance of the work for the field. As a whole, the series will provide a valuable resource both for newcomers to the research fields described, and for other scientists seeking detailed background information on special questions. Finally, it provides an accredited documentation of the valuable contributions made by today's younger generation of scientists.

Theses are accepted into the series by invited nomination only and must fulfill all of the following criteria

- They must be written in good English.
- The topic should fall within the confines of Chemistry, Physics, Earth Sciences, Engineering and related interdisciplinary fields such as Materials, Nanoscience, Chemical Engineering, Complex Systems and Biophysics.
- The work reported in the thesis must represent a significant scientific advance.
- If the thesis includes previously published material, permission to reproduce this must be gained from the respective copyright holder.
- They must have been examined and passed during the 12 months prior to nomination.
- Each thesis should include a foreword by the supervisor outlining the significance of its content.
- The theses should have a clearly defined structure including an introduction accessible to scientists not expert in that particular field.

More information about this series at http://www.springer.com/series/8790

Kosuke Ohsawa

Total Synthesis of Thielocin B1 as a Protein–Protein Interaction Inhibitor of PAC3 Homodimer

Doctoral Thesis accepted by
Tohoku University, Sendai, Japan

Springer

Author
Dr. Kosuke Ohsawa
Tohoku University
Sendai
Japan

Supervisor
Prof. Takayuki Doi
Tohoku University
Sendai
Japan

ISSN 2190-5053 ISSN 2190-5061 (electronic)
Springer Theses
ISBN 978-4-431-55446-2 ISBN 978-4-431-55447-9 (eBook)
DOI 10.1007/978-4-431-55447-9

Library of Congress Control Number: 2014960203

Springer Tokyo Heidelberg New York Dordrecht London

Printed on acid-free paper

Springer Japan KK is part of Springer Science+Business Media (www.springer.com)

Parts of this thesis have been published in the following journal articles:

1. "A Direct and Mild Formylation Method for Substituted Benzenes Utilizing Dichloromethyl Methyl Ether - Silver Trifluoromethanesulfonate" Ohsawa, K.; Yoshida, M.; Doi, T. *J. Org. Chem.* **2013**, *78*, 3438–3444. *Reproduces with permission*
2. "Total Synthesis and Characterization of Thielocin B1 as a Protein-Protein Interaction Inhibitor of PAC3 Homodimer" Doi, T.; Yoshida, M.; Ohsawa, K.; Shin-ya, K.; Takagi, M.; Uekusa, Y.; Yamaguchi, T.; Kato, K.; Hirokawa, T.; Natsume, T. *Chem. Sci.* **2014**, *5*, 1860–1868. *Reproduces with permission*

Supervisor's Foreword

It is my pleasure to introduce Dr. Kosuke Ohsawa's study for publication in the Springer Thesis series as an outstanding original work from one of the world's top universities. Dr. Ohsawa was born in Japan in 1987. After graduation in 2009 from the Department of Pharmaceutical Sciences, Tohoku University, he received a license to practice pharmacy. He studied Organic Chemistry in the Ph.D. course at the Graduate School of Pharmaceutical Sciences of Tohoku University and received his Ph.D. in 2014. He moved to Dortmund, Germany, as a postdoctoral fellow under the guidance of Prof. Dr. Herbert Waldmann in Max-Planck Institute of Molecular Physiology.

Protein–protein interactions (PPIs) have recently attracted attention as novel therapeutic targets because they play vital roles in numerous cellular functions. However, conventional methodology based on lock-and-key theory is usually difficult to apply for large and flat interfaces of PPIs. Dr. Ohsawa focused on thielocin B1, which was found to be a PPI inhibitor of a proteasome assembling chaperone (PAC) 3 homodimer, and he has achieved the first total synthesis of thielocin B1 and its molecular probe. In the process, he developed the synthetic methodology for highly functionalized aromatic compounds, e.g., the synthesis of unique 2,2′,6,6′-tetrasubstituted diphenyl ether and the formylation of aromatic compounds at sterically hindered position. Compounds that he synthesized were utilized for NMR studies to validate the interaction model of a thielocin B1/PAC3 complex, and further in silico docking study suggests that thielocin B1 promotes the dissociation to monomeric PAC3 after approaching PAC3 homodimer.

I hope that Dr. Ohsawa's Ph.D. thesis will help many readers to reconfirm the potential of complicated natural products as drug seeds.

Sendai, October 2014 Prof. Takayuki Doi

Acknowledgments

I express my sincere and wholehearted appreciation to Prof. Dr. Takayuki Doi (Graduate School of Pharmaceutical Sciences, Tohoku University) for constructive discussions, constant encouragement, and kind guidance as a chemist during this study.

I would like to show my greatest appreciation to Dr. Masahito Yoshida (Graduate School of Pharmaceutical Sciences, Tohoku University) for tremendous support and extensive discussions.

I am deeply grateful to Dr. Hirokazu Tsukamoto (Graduate School of Pharmaceutical Sciences, Tohoku University), Dr. Yuichi Masuda (Graduate School of Pharmaceutical Sciences, Tohoku University), Prof. Dr. Ko Hiroya (Musashino University), and Dr. Kiyofumi Inamoto (Mukogawa Women's University) for valuable advice and helpful guidance.

I also would like to express my gratitude to Prof. Dr. Masahiko Yamaguchi (Graduate School of Pharmaceutical Sciences, Tohoku University) and Dr. Naoki Kano (Graduate School of Pharmaceutical Sciences, Tohoku University) for invaluable comments on the revision of my original manuscript.

I thank Dr. Kazuo Shin-ya (National Institute of Advanced Industrial Science and Technology), Dr. Tohru Natsume (National Institute of Advanced Industrial Science and Technology), and Prof. Dr. Motoki Takagi (Fukushima Medical University) for evaluating the biological activity of synthesized thielocin B1 for PAC3 homodimer. I also appreciate Dr. Takatsugu Hirokawa (National Institute of Advanced Industrial Science and Technology) for performing in silico docking studies. I acknowledge Prof. Dr. Koichi Kato (Institute for Molecular Science, National Institutes of Natural Science), Dr. Yoshinori Uekusa (National Institute of Health Sciences), and Dr. Takumi Yamaguchi (Institute for Molecular Science, National Institutes of Natural Science) for conducting NMR experiments.

I appreciate Ms. Yuki Sato, Ms. Yukie Ogiyanagi, and Ms. Mieko Inoue for their office work. I am also grateful to all my colleagues at the Doi Laboratory, Graduate School of Pharmaceutical Sciences of Tohoku University, for their valuable discussion and comments.

Finally, I express my gratitude to my parents, Ken-ichi and Eiko Ohsawa, and to all members of my family for their constant encouragement.

Contents

Chapter 1
Introduction

1.1 Introduction

Natural products often exhibit a broad range of interesting biological activities, and numerous natural products have been used as therapeutic agents in traditional folk medicines for many centuries. However, the core structures of many natural products are incredibly complicated, making them unsuitable for the rapid elucidation of structure-activity relationships (SARs) in conventional drug discovery programs, and many natural products are therefore used as tool compounds for target validation in synthetic organic chemistry. Despite the difficulties associated with their synthesis, the importance of natural products as lead compounds in drug discovery should not be overlooked because they can sometimes offer unique therapeutic mechanisms of action compared with the small drug-like molecules found in most conventional high-throughput screening (HTS) libraries. As part of our ongoing research into the development of new drug candidates from natural products, we recently identified thielocin B1 as a novel inhibitor of the protein–protein interaction (PPI) between PAC3 homodimer following the HTS of a natural products library. Several examples have been provided in this chapter to highlight the potential of natural products as lead compounds for the development of therapeutic agents for undeveloped drug targets.

1.2 Natural Products as Drug Seeds

For many centuries, numerous cultures have sought to identify therapeutic agents for the treatment of diseases from natural sources such as plants. Subsequent developments in analytical and synthetic chemistry have allowed for the isolation and identification of the bioactive compounds responsible for the therapeutic effects of these natural materials and their clinical application in modern-day medicine.

K. Ohsawa, *Total Synthesis of Thielocin B1 as a Protein–Protein Interaction Inhibitor of PAC3 Homodimer*, Springer Theses, DOI 10.1007/978-4-431-55447-9_1

The first example of this type is the antipyretic analgesic aspirin (**1**), which was developed in 1899 by Bayer AG following on from the isolation of salicylic acid (**2**) from the bark of the willow tree. Many of the drugs currently used in modern medicine were derived from natural products, including the analgesic morphine (**3**), the anticancer drug taxol (**4**) and the antibacterial agent penicillin G (**5**). Marine natural products have recently emerged as an exciting area of research in drug discovery because of the unique molecular architectures and interesting biological properties of compounds such as nakadomarin A (**6**), cortistatin A (**7**) and norzoanthamine (**8**) (Fig. 1.1).

Although many natural products exhibit potent biological activities, they are often overlooked as lead compounds in drug discovery programs. One of the key reasons for not using natural products as lead compounds in drug discovery is that the supply of these compounds from natural sources is often limited, and complex organic syntheses are therefore required to provide access to these materials and their derivatives for SAR studies. Furthermore, robust and efficient manufacturing routes would need to be developed to provide access to the large quantities of pure bioactive compounds needed to progress a natural product through clinical trials and onto the market, which could have significant cost, safety and environmental load implications. A good example of this particular problem is provided by taxol (**4**), which has been the subject of numerous total syntheses [1–9] and one formal synthesis [10]. Notably, however, none of these synthetic routes has been applied to the process-scale synthesis of **4** because too many steps are required and these syntheses are not amenable to large-scale manufacture. Instead, compound **4** is manufactured by a semisynthetic route from 10-deacethylbaccatin III (**9**) [11], which can be readily isolated from the leaves of *Taxus baccata* (Scheme 1.1).

Although low-molecule-weight compounds, which can be readily synthesized over a short number of steps using reliable reactions, are the current mainstream of drug discovery, the probability of new drugs being generated from these

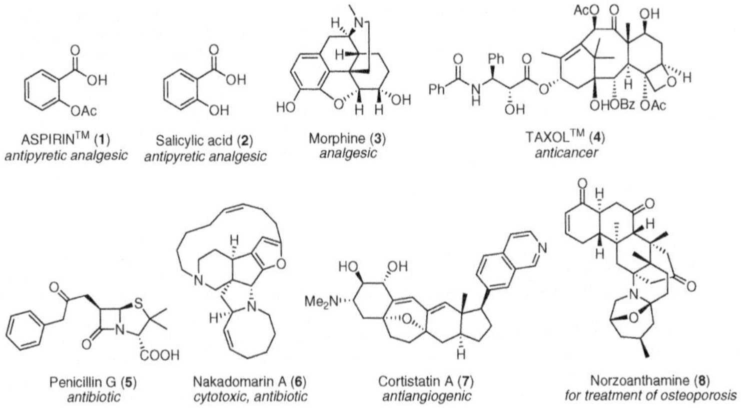

ASPIRIN™ (**1**) *antipyretic analgesic*
Salicylic acid (**2**) *antipyretic analgesic*
Morphine (**3**) *analgesic*
TAXOL™ (**4**) *anticancer*
Penicillin G (**5**) *antibiotic*
Nakadomarin A (**6**) *cytotoxic, antibiotic*
Cortistatin A (**7**) *antiangiogenic*
Norzoanthamine (**8**) *for treatment of osteoporosis*

Fig. 1.1 Bioactive compounds derived from natural products

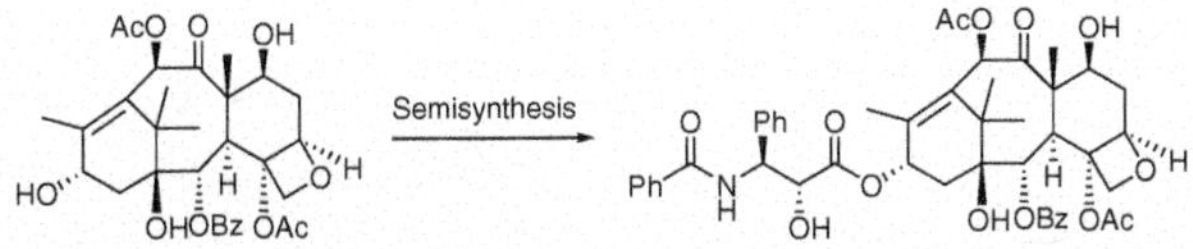

10-Deacetylbaccatin III (9) → Semisynthesis → TAXOL™ (4)

Scheme 1.1 Semisynthesis of taxol (**4**)

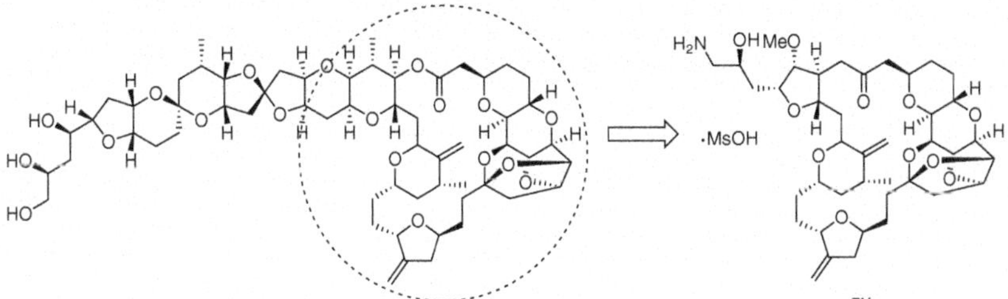

Halichondrin B (11) ⇒ HALAVEN™ (10) (Eribulin mesylate)

Fig. 1.2 Natural product-inspired drug design of halaven (**10**)

compounds has not been improved even with developments in methodology such as combinatorial chemistry and HTS. The importance of natural products as drug leads is therefore reconfirmed. In 2010, halaven (**10**) was marketed by Esai Co. for the treatment of metastatic breast cancer. Compound **10** is a macrocyclic ketone analogue of halichondrin B (**11**), which was isolated from marine sponge [12] (Fig. 1.2). The fact that **10** is currently manufactured using a fully organic synthesis [13] highlights the usability of natural products as drug leads and high-quality tools in modern synthetic organic chemistry.

1.3 Protein–Protein Interactions

PPIs play a vital role in numerous cellular processes, including cell growth, DNA replication and transcriptional activation [14]. Proteomics has been used to show network of PPIs with the transition from human-genomic era to the post-genomic era, and PPIs such as p53/MDM2 [15] and members of the Bcl-2 family [16] are considered to be potential drug discovery targets. However, there are currently no inhibitors targeting intracellular PPIs on the market. Given that the specific areas of the proteins involved in PPIs are quite different from those of receptors, it is difficult to use conventional drug discovery approaches based on the lock and key theory proposed by Paul Ehrlich to disrupt these interactions (Fig. 1.3) [17]. Typical receptors, such as those found on GPCRs and enzymes, usually interact with their natural ligands through a series of noncovalent interactions in their ligand-binding pocket, which provide a large contact area. These binding sites usually project deep

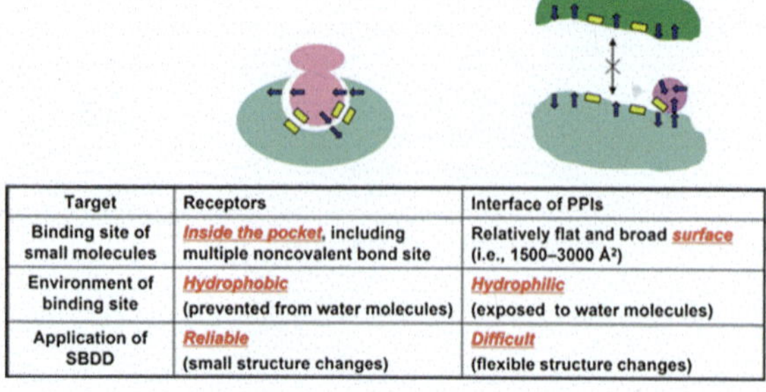

Target	Receptors	Interface of PPIs
Binding site of small molecules	Inside the pocket, including multiple noncovalent bond site	Relatively flat and broad surface (i.e., 1500–3000 Å²)
Environment of binding site	Hydrophobic (prevented from water molecules)	Hydrophilic (exposed to water molecules)
Application of SBDD	Reliable (small structure changes)	Difficult (flexible structure changes)

Fig. 1.3 Differences between receptors and the interfaces of PPI as drug targets

inside the protein and are consequently cut off from external water molecules and largely unaffected by structural changes elsewhere in the protein. Binding sites of this type are therefore compatible with structure-based drug design (SBDD), which can be used to design drug-like compounds based on the natural ligands. In contrast, the areas of proteins involved in PPIs are relatively flat and much larger (i.e., 1,500–3,000 Å^2) [18] than those involving the modulation of a target with a small molecule (i.e., 300–1,000 Å^2) [19]. The large surface areas involved in PPIs make it difficult for small compounds to make meaningful and high-affinity interactions at the interface of PPIs through multiple noncovalent bonds. The use of SBDD to develop predictive models for the interaction of small molecules at the interface of PPIs can be also difficult because the structure of the interface can change significantly depending on the nature of the external environment and the PPI itself. Furthermore, there are two possible mechanisms for inhibiting PPIs, whereby an inhibitor binds to the target protein before or after dissociation (i.e., pre-dissociation independent and pre-dissociation dependent mechanisms, respectively) [20] (Fig. 1.4). The complexity of these mechanisms makes it especially difficult to use SBDD to design inhibitors of PPIs.

Recent studies have suggested that small-molecule inhibitors would not need to cover the entire interface of PPIs but only interact with high-affinity regions, known as "hot spots" [21]. Hot spots mainly consist of hydrophobic interactions, which are presumably designed to avoid the influence of external water. Furthermore, it has been revealed that there are generally small grooves and indentations at the interfaces of PPIs [22]. If there are hot spots close to these grooves, then it is possible

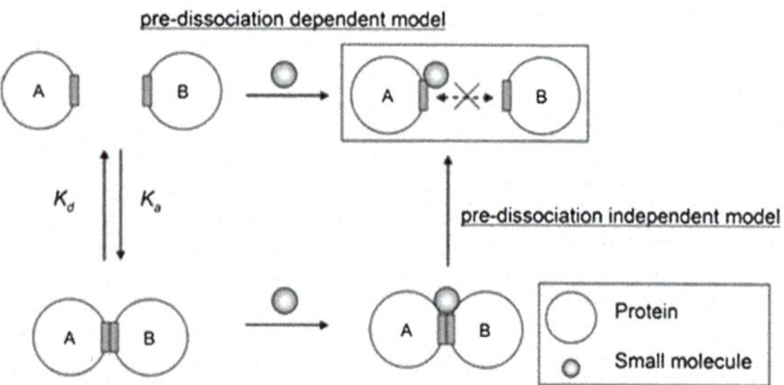

Fig. 1.4 Two possible mechanisms for the inhibition of PPIs

Fig. 1.5 p53 mimic **12** as a p53/MDM2 inhibitor

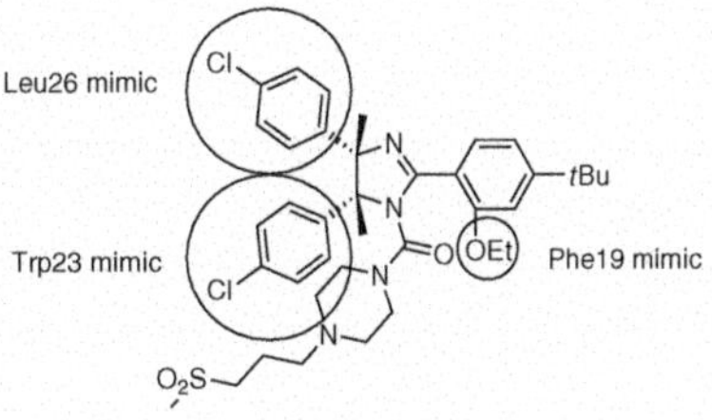

that a small molecule could bind to these sites and behave as an inhibitor of PPIs. In fact, the p53/MDM2 and Bcl-X_L/Bak complexes mentioned above are two of the most well-researched PPI targets to possess hot spots that appear as wide and shallow grooves. RG7112 (**12**) was recently developed by Hoffmann-La Roche as an inhibitor of the p53/MDM2 PPI [23]. The ethoxy and two 4-chlorophenyl groups of **12** imitate the Phe19, Trp 23 and Leu 26 residues of p53, respectively, to interact with a groove on the surface of MDM2 through a series of hydrophobic interactions (Fig. 1.5). In a separate example, Abbott developed ABT-737 (**13**) as a mimic of the BH3 domain in Bak to inhibit the PPI between Bcl-X_L and BH3 [24]. The oral bioavailability of compound **13** and its affinity for Bcl-X_L were improved by reducing one of its phenyl rings to a cyclohexene and introducing a pyrrolopyridine ring, respectively (Fig. 1.6). ABT-263 (**14**) [25] and ABT-199 (**15**) [26] are currently being evaluated in clinical trials for the treatment of chronic lymphocytic leukemia.

Based on the success of these compounds, it seems clear that the lock and key theory associated with traditional drug discovery can be applied to PPIs if the hot spots on the surfaces of the proteins are close to the grooves and indentation at the interface of the PPI of interest. In this case, planar compounds would be better than those with a high level of three-dimensional character to provide high-affinity hydrophobic interactions with the wide and shallow grooves on the surfaces of

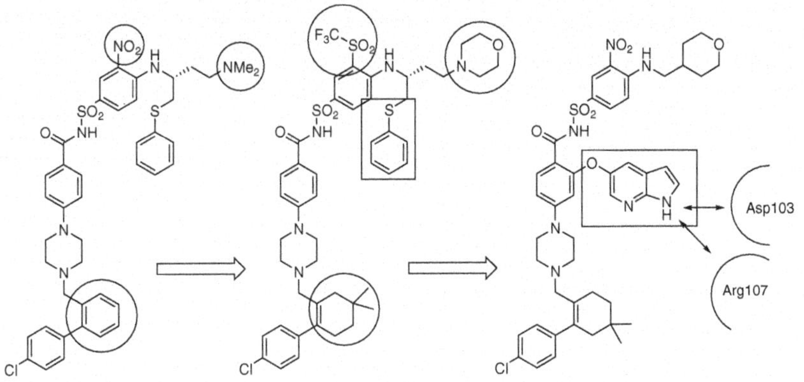

Fig. 1.6 BH3 domain mimics **13–15** as Bcl-X_L/BH3 inhibitors

PPIs. However, conventional strategies are not always effective for the development of PPI inhibitors because the molecular weights of many inhibitors tend to increase significantly using large numbers of hydrophobic functional groups such as aromatic rings.

1.4 Evaluation of PPI Inhibitors with High-Throughput Screening

Given that the unique characteristics of the interfaces of PPIs make it difficult to find suitable drug leads as PPI inhibitors, the exhaustive evaluation of large libraries of inhibitors by HTS represents one of the most efficient strategies for identifying new leads. In this way, the activities of the compounds can be investigated without being limited to known binding modes, although it is important to mention that conventional libraries composed of small drug-like molecules are not best suited to the identification of leads compounds because they occupy such small surface areas and are generally made up of similar structures.

In 2009, Shin-ya et al. [27] conducted a HTS campaign of over 100,000 diverse samples from crude metabolites and isolated natural products to identify PPI inhibitors. The unique structures found in natural products can occupy extensive areas of chemical space, and it was therefore expected that libraries of this type would provide higher hit rates than conventional small-molecule libraries against PPIs. Three PPIs, including TCF7/β-catenin, proteasome-assembling chaperone 1 (PAC1)/PAC2 and PAC3 homodimer, have been identified as playing important roles in the existence of cancer cells and these PPIs were selected as target proteins. As a result of screening with monomeric Kusabira-Green fluorescent protein in vitro [28, 29], selective hits were confirmed for the last two of three targets although the hit rates were less than 0.01 %. Among them, thielocin B1 (**16**) was found to be a potent and selective inhibitor of PAC3 homodimer ($IC_{50} = 0.020\ \mu M$) [27], as was JBIR-22 (**17**) ($IC_{50} = 0.20\ \mu M$) [30] (Fig. 1.7). The fact that **16** recognized a specific structure on the surface of PAC3, rather than exhibiting nonspecific inhibition, was very interesting in terms of elucidating its mode of interaction.

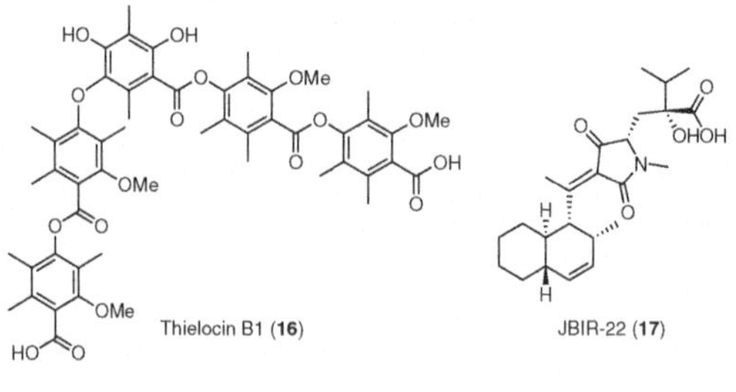

Thielocin B1 (**16**) JBIR-22 (**17**)

Fig. 1.7 Natural products as selective inhibitors of PAC3 homodimer

1.5 Proteasome-Assembling Chaperone

The degradation of polyubiquitinated proteins is promoted by 26S proteasome. Given that this proteasome is specifically activated in cancer cells, an inhibitor of its activity would induce apoptosis and could therefore be used as a potential treatment for cancer. Many activators are required to supply 20S proteasome, which is a main component of 26S proteasome. Among them, PAC3 promotes the association of specific molecules in cooperation with PAC1/PAC2 to provide the α ring. After the formation of the β ring, 20S proteasome undergoes a dimerization process with half of the proteasome (Fig. 1.8) [31, 32]. It has recently been reported that PAC4 interacts with PAC3, and that the resulting complex acts as a promoter [33–35]. Yashiroda et al. [36] proposed that PAC3 forms a homodimer based on the elucidation of its crystal structure, and that monomeric PAC3 can therefore only interact with PAC4 following the dissociation of PAC3 homodimer.

The binding of bortezomib (**18**) to the active site of the 26S proteasome has been used as a therapeutic strategy for the treatment of multiple myeloma [37] (Fig. 1.9). In contrast, compound **16** interacts with PAC3 homodimer, which is an activator of the 20S proteasome and should be considered a novel drug lead for the development of proteasome inhibitors.

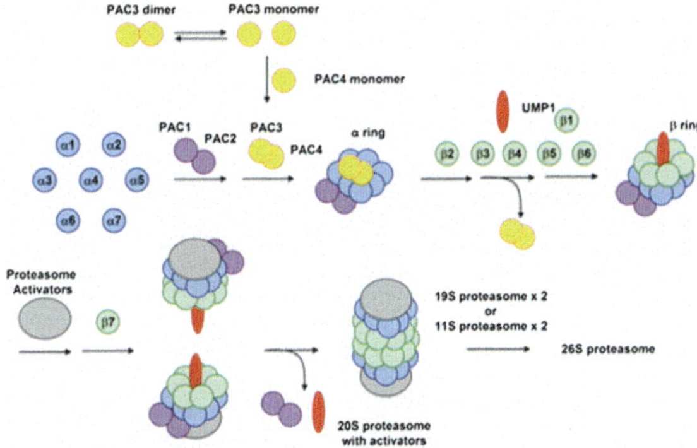

Fig. 1.8 Formation of 20S proteasome by PAC

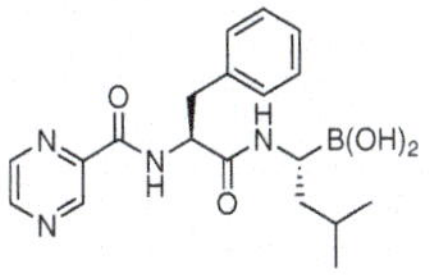

Bortezomib (**18**)

Fig. 1.9 Structure of the 26S proteasome inhibitor bortezomib (**18**)

1.6 Thielocin B1

Thielocin B1 (**16**) was isolated from the fermentation broth of *Thielavia terricola* RF-143 in 1995, and subsequently found to be a potent inhibitor for phospholipase A2 [38]. The HTS of a natural product library to identify novel PPI inhibitors revealed that **16** strongly inhibited the formation of PAC3 homodimer [27]. Surprisingly, however, the similar compounds thielocins A1β (**19**) and B3 (**20**) did not inhibit the formation of PAC3 homodimer ($IC_{50} > 250$ μL), which indicated that the central core in **16** played an important role in its activity towards PAC3 homodimer because all three of these compounds possessed the same side wings (Fig. 1.10).

Compound **16** consisted of five fully substituted phenyl rings, which are connected through one ether and three ester linkages. In particular, the electron-rich 2,2′,6,6′-tetrasubstituted diphenyl ether moiety of this compound is a rare structural feature that is present in very few natural and unnatural products. The syntheses of **19** [39, 40], **20** [41] and various derivatives of **20** [42] have been reported. In contrast, the total synthesis of **16** has not been achieved because the 2,2′,6,6′-tetrasubstituted diphenyl ether moiety is very difficult to form. The supply of **16** from natural sources is limited, and it is therefore necessary to establish an efficient synthetic route to this compound so that its inhibition mechanism can be fully elucidated.

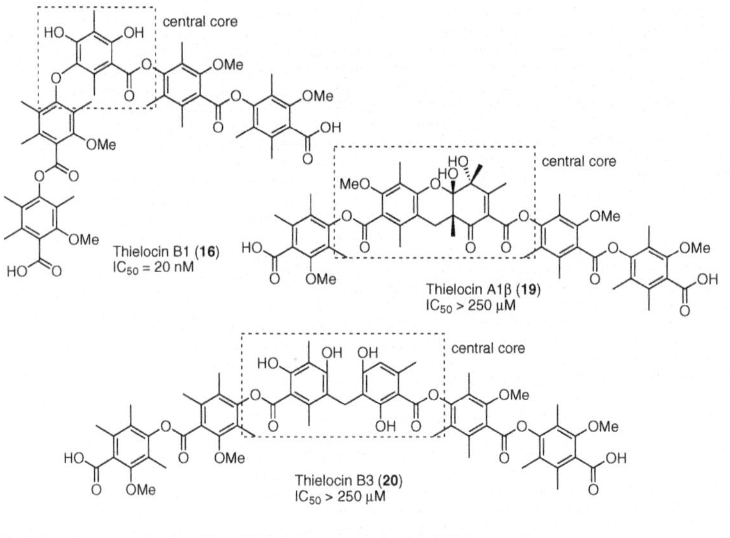

Fig. 1.10 IC_{50} values of specific thielocins towards PAC3 homodimer

1.7 Preliminary Investigation to Elucidate the Mechanism of Inhibition of PAC3 Homodimer by Thielocin B1

1.7.1 In Silico Docking Studies and Molecular Dynamics Simulation Experiments

Gilde docking [43] studies were performed by Hirokawa [44] for active **16** and inactive **20** at the interface of PAC3 (PDB code: 2Z5E_A) [36] to determine the mode of inhibition of compound **16**. The irreversible interaction of a small molecule at the interface of a PPI can lead to the formation of a strong small molecule/protein complex. The top 100 energetically favored poses of **16** and **20** were separately overlapped on the X-ray crystal structure of PAC3. Interestingly, nearly all of the poses for compound **16** indicated that it was suitably shaped to interact with the hill-like β-sheet structure on the surface of PAC3 (Fig. 1.11a). In contrast, compound **20** provided a variety of scattered poses, which implied that it was making nonspecific interactions with the surface of the protein (Fig. 1.11b). The lower docking scores of **16** compared with **20** also supported the results (Fig. 1.11c). Consideration of the central cores of **16** and **20** revealed that the diphenylmethane moiety in **20** was much more flexible than the rigid phenyl 3-aryloxybenzoate moiety in **16** (see Fig. 1.10). It is surprising that the well-defined angle created by the central core in **16** recognizes and strongly binds to the hill-like structure on the surface of PAC3 protein, despite there being no obvious cavities.

Molecular dynamics (MD) simulations were also performed to examine the factors responsible for the high affinity of the **16**/PAC3 complex [44]. The results of protein-ligand interaction fingerprint analysis revealed an overall abundance of greater than 40 % among the last 10 ns MD simulation trajectories for five residues on PAC3, including Ser30, His31, Lys65, Leu67 and Asp71 (Fig. 1.12a), and a representative structure of the **16**/PAC3 complex is shown in Fig. 1.12b. From the X-ray crystal structure of PAC3 homodimer, it has been confirmed that Ser30 and His31 interact with Gly40′, Lys41′ and Met42′ (Fig. 1.12c). The carboxyl group in **16** forms hydrogen bonds with Ser30 and His31 (97 and 86 %, with the side-chain atoms acting as hydrogen bond donors; 91 and 61 %, surface contact interactions, respectively) to interrupt the connection. Furthermore, the dimethylbenzene ring in **16** interacts with the methylene moiety of Lys65 (96 % surface contact interactions). This hydrophobic interaction induces the amino group in Lys65 to approach Glu49 to form a new hydrogen bond, which causes the conformation of PAC3 to change. The central core of **16** remains in close proximity to Leu67 through a series of hydrophobic interactions (89 % surface contact interactions) and prevents this residue from interacting with Leu67′. Given that the leucine-rich area (Leu45, Leu67, Leu68 and Leu96) of the β-sheet structure of PAC3 forms a hydrophobic cluster involved in the formation of a strong PPI (Fig. 1.12d), this interruption of this interaction would have an adverse impact on the formation of the homodimer. One of the methoxy groups in **16** most likely interacts with the Asp71 residue (47 % surface contact interactions) to support the **16**/PAC3 complex.

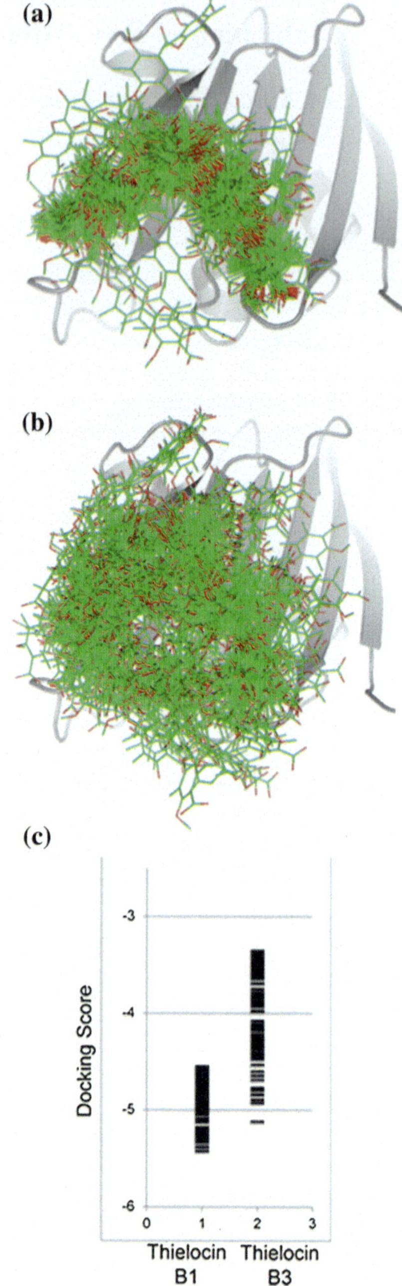

Fig. 1.11 Overlap of the top 100 energetically favored poses of thielocins (carbons in *green*) at PAC3 interface. **a** Thielocin B1 (**16**). **b** Thielocin B3 (**20**). **c** Docking score distribution for the **16**/PAC3 and **20**/PAC3 complexes

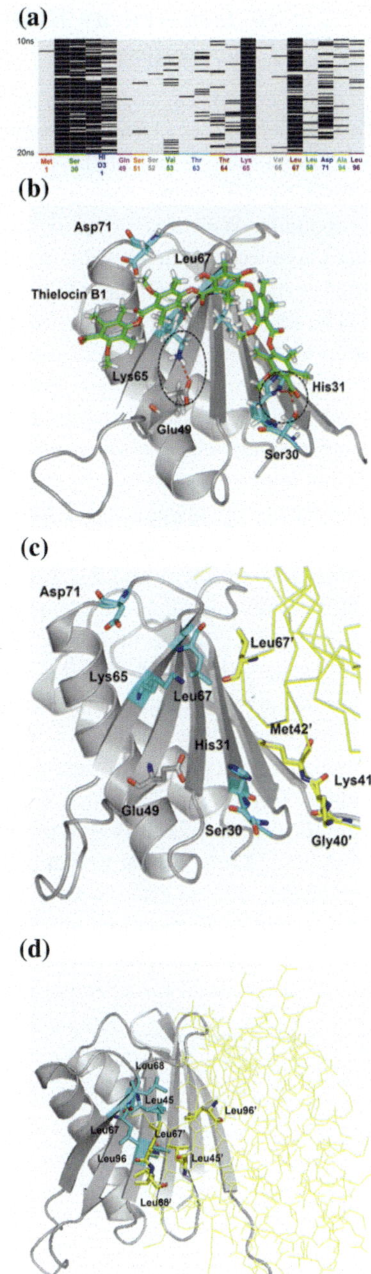

Fig. 1.12 **a** Barcode view of the protein-ligand interaction fingerprint: Met1: Surf; Ser30: ChDon, Surf; His31: ChDon, Surf; Glu49: Surf; Ser51: Surf; Ser52: Surf; Val53: Surf; Thr63: ChDon, Surf; Thr64: ChDon, Surf; Lys65: ChDon, Surf; Val66: BkDon, Surf; Leu67: Surf; Leu68: Surf; Asp71: Surf; Ala94: Surf; Leu96: Surf. ChDon = side-chain atoms act as hydrogen bond donors. BkDon = backbone atoms act as hydrogen bond donors. Surf = surface contract interactions. **b** Representative structure of the **16**/PAC3 complex. Five interacting residues (Ser30, His31, Lys65, Leu67 and Asp71) identified by protein-ligand interaction fingerprint analysis with >40 % overall abundance among the last 10 ns of the MD simulation trajectories, which are shown as carbons in cyan. **c** The X-ray crystal structure of PAC3 homodimer (PDB code: 2Z5E). Ser30-His31/Gly40′-Met42′ and Leu67/Leu67′ were found to be at the interface. **d** The hydrophobic cluster generated by Leu45, Leu67, Leu68 and Leu96. These side chains overhang from the surface

1.7.2 Structure-Activity Relationships of Several Thielocin B1 Derivatives

Various derivatives of **16** have been synthesized by Yoshida, and their bioactivities as PPI inhibitors elucidated. As mentioned above, the boomerang-like structure defined by the central core in **16** is important to the inhibitory activity of these compounds. With this in mind, several researchers have investigated the use of alternative core structures to develop a deeper understanding of the important features of this central core (Fig. 1.13).

The results of assays designed to assess the inhibitory activities of derivatives based on **21** and side wing **22** towards the PA1/PAC2 and PAC3 homodimer PPIs are summarized in Table 1.1. Based on the speed with which they could be prepared, the symmetric dibenzylic ether and diester groups were initially selected as suitable candidates for the central core. Although diethers **21a** and **21c** exhibit distinct inhibitory activity towards PAC3 homodimer at 10 μM, they exhibited very little selectivity towards the PAC1/PAC2 complex (Table 1.1, entries 2 and 4). Diester **21e** and **21g** also inhibited both of these complexes in a nonspecific manner (Table 1.2, entries 6 and 8). An evaluation of compounds **21e** and **21g** as PPI inhibitors revealed that the presence of a free hydroxyl group on the central core was not critical to the inhibitory activity of these compounds. In contrast, the di-MOM esters **21b**, **21d** and **21f**, where MOM groups were used to protect the terminal carboxyl groups, did not exhibit any bioactivity towards the PAC1/PAC2 or PAC3 homodimer (Table 1.2, entries 3, 5 and 7). These results suggested that the terminal carboxyl groups were essential to the inhibitory activity of these compounds. Compound **22**, which is a substructure of **16**, was not effective as a PPI inhibitor (Table 1.2, entry 9). Among these derivatives, it should be noted that the structure of **21e** is similar to that of **16**, except that one of the ether bonds in **16** has been changed to an ester bond in **21e**. The difference in the inhibitory activities of these compounds therefore highlights the importance of this ether group to the activity of **16** and therefore indicates that similar bonding styles to those found in **16** will be required for selective interaction with PAC3 homodimer.

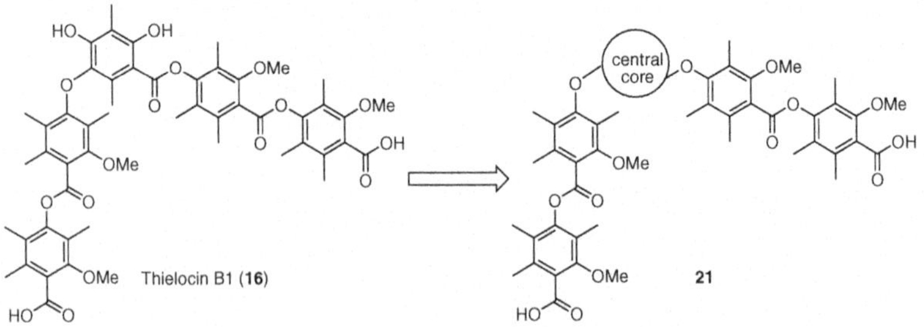

Fig. 1.13 Design of thielocin B1 derivatives **21**

Table 1.1 Evaluation of derivatives of **21** and side wing **22** as PPI inhibitors

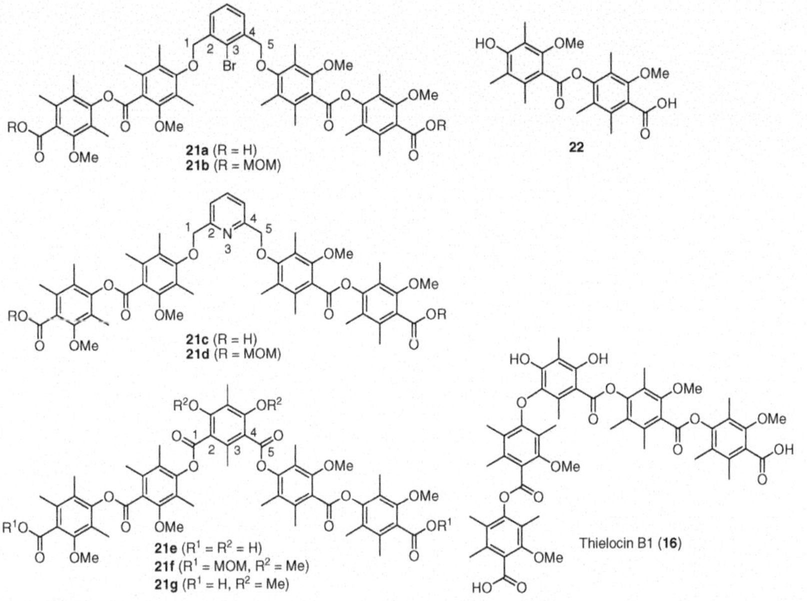

21a (R = H)
21b (R = MOM)

22

21c (R = H)
21d (R = MOM)

21e ($R^1 = R^2 = H$)
21f (R^1 = MOM, R^2 = Me)
21g (R^1 = H, R^2 = Me)

Thielocin B1 (**16**)

		Inhibition rate for PAC3/PAC3[a] (%)			Inhibition rate for PAC1/PAC2[b] (%)		
Entry	Sample	10 μM[c]	1 μM[c]	0.1 μM[c]	10 μM[c]	1 μM[c]	0.1 μM[c]
1	**16**	–	94	85	–	9	9
2	**21a**	88	43	18	40	5	9
3	**21b**	0	0	0	17	0	3
4	**21c**	86	38	26	57	5	3
5	**21d**	0	0	9	6	0	3
6	**21e**	94	17	4	69	23	21
7	**21f**	2	5	12	1	0	3
8	**21g**	85	49	29	38	2	9
9	**22**	21	9	11	11	0	3

[a] mKG_N-PAC3 (0.010 μM), mKG_C-PAC3 (0.025 μM)
[b] PAC1-mKG_N (0.079 μM), mKG_C-PAC2 (0.12 μM)
[c] Concentration of samples in DMSO

Based on these results, several asymmetric derivatives bearing the same bonding linkers as those found in **16** were investigated, as shown in Table 1.2. However, these new compounds, **21h–j**, which were synthesized using an Ullmann coupling reaction, exhibited strong and nonspecific inhibition towards both the PA1/PAC2 and PAC3 homodimer at 10 μM (Table 1.2, entries 2–4). These results therefore indicated that it is not only important to imitate the bonding style in **16** but it is also important to imitate the hydrophobic interaction of the methyl group on the central

Table 1.2 Evaluation of asymmetric derivatives of **21** as PPI inhibitors

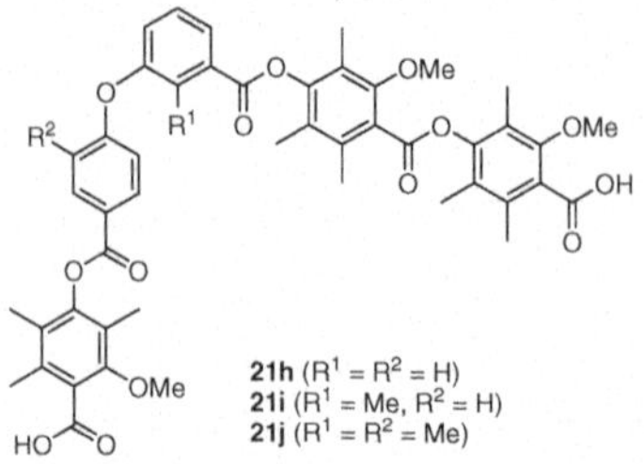

21h ($R^1 = R^2 = H$)
21i ($R^1 = Me$, $R^2 = H$)
21j ($R^1 = R^2 = Me$)

		Inhibition rate for PAC3/PAC3[a] (%)			Inhibition rate for PAC1/PAC2[b] (%)		
Entry	Sample	10 μM[c]	1 μM[c]	0.1 μM[c]	10 μM[c]	1 μM[c]	0.1 μM[c]
1	**16**	–	94	85	–	9	9
2	**21h**	99	37	7	98	35	9
4	**21i**	98	40	4	91	37	15
6	**21j**	99	34	19	89	31	1

[a] mKG_N-PAC3 (0.010 μM), mKG_C-PAC3 (0.025 μM)
[b] PAC1-mKG_N (0.079 μM), mKG_C-PAC2 (0.12 μM)
[c] Concentration of samples in DMSO

core to provide selective inhibition. It is interesting to note that the unique structure of the natural product **16** covers an area of chemical space that is not usually supported by the synthetic compounds found in traditional HTS libraries.

1.8 Synthesis of *Ortho*-substituted Aromatic Compounds

1.8.1 Synthesis of Multisubstituted Diphenyl Ethers

The diphenyl ether moiety is found in a wide range of bioactive compounds, such as vancomycin (**23**) and bouvardin (**24**), and is an important structural class with numerous applications in the life sciences. Furthermore, the diphenyl ether moiety has been used as an important component of several polymeric materials, including polyetherketone (**25**) and polyetheretherketone (**26**) (Fig. 1.14).

In 1903, Ullmann et al. [45–48] reported the Cu-mediated coupling reaction of phenols **27a** with aryl halides **28** to give the corresponding diphenyl ethers **29**. Although this reaction was the first reported example of a methodology for the formation of a C(sp^2)–O bond, its scope was limited by its requirement for stoichiometric quantities of Cu and high reaction temperatures. During the 1970s, it was established that Ni and Pd could be used as catalysts for the formation of C(sp^2)–C(sp^2) bonds. Despite significant advances in this area of chemistry during

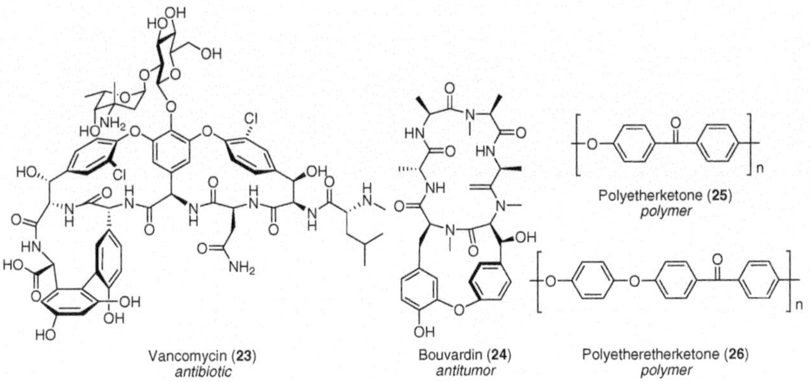

Fig. 1.14 Natural and unnatural compounds containing a diphenyl ether moiety

the decades that followed, the transition metal-mediated formation of C(sp^2)–O bonds continued to represent a significant challenge to synthetic organic chemistry. The main problem with this transformation is that a Pd–O bond has greater ion-bonding properties than a Pd–C(sp^2) bond, which leads to a much slower reductive elimination step for reactions involving the formation of a C(sp^2)–O bond [49]. The breakthrough required to solve this serious problem was achieved through the development of novel ligands for this transformation. During the 1990s, Buchwald et al. [50] and Hartwig et al. [51] reported the development of Pd-catalyzed coupling reactions for the synthesis of diphenyl ethers. To achieve this transformation, Buchwald and Hartwig developed bulky alkylphosphines **30** and **31** as ligands, which enhanced the rate of the reductive elimination step as a consequence of their wide cone angles. Furthermore, these reactions only required a catalytic amount of Cu to facilitate the Ullmann coupling in the presence of **32** [52]. Following on from these advances, the transition metal-catalyzed synthesis of diphenyl ether moieties is currently one of the most heavily researched areas of chemistry involving coupling reactions [53–55] (Scheme 1.2).

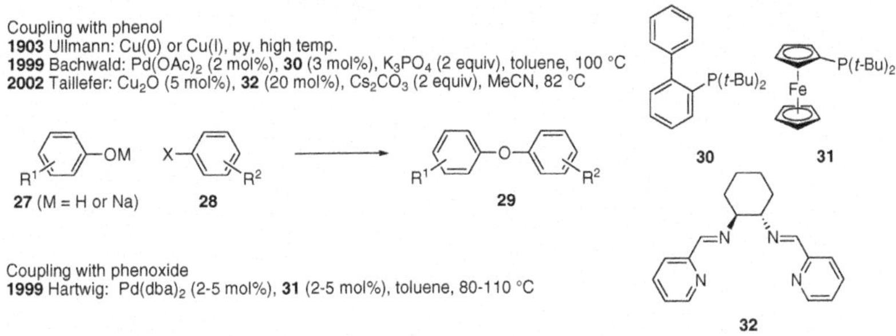

Scheme 1.2 Synthesis of diphenyl ether **29** using transition metals

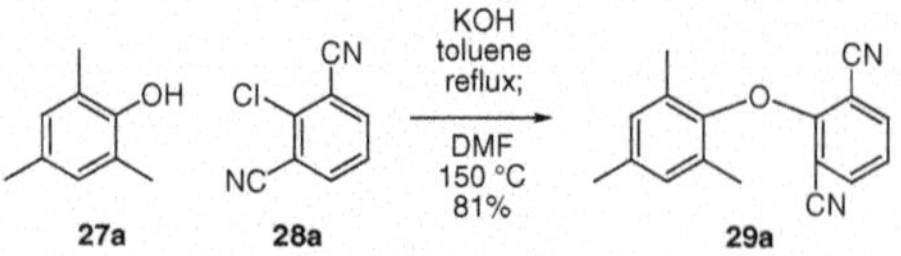

Scheme 1.3 Synthesis of diphenyl ether **29a** using an S_NAr reaction

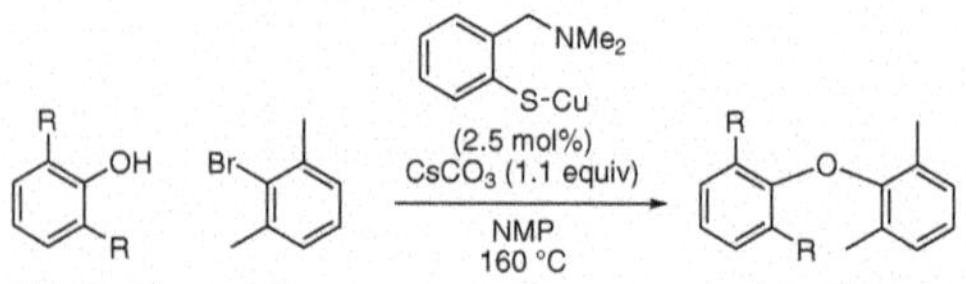

Scheme 1.4 Synthesis of diphenyl ethers **29b** and **29c** using a coupling reaction

Bulky alkylphosphine ligands also promote the oxidative addition steps of numerous coupling reactions because they are highly electron-donating. Although ligands of this type are often used for the synthesis of 2,2′,6,6′-tetrasubstituted biphenyl groups, it has been reported that these ligands still find it difficult to form 2,2′6-tri- and 2,2′,6,6′-tetrasubstituted diphenyl ethers. In these cases, a phenoxide group can be eliminated from the Pd catalyst prior to the reductive elimination step to avoid steric hindrance. Alternatively, aryl halides bearing an electron-withdrawing group can undergo an S_NAr reaction to afford diphenyl ethers (Scheme 1.3) [56]. In contrast, there have been no reports in the literature to date pertaining to the efficient synthesis of electron-rich 2,2′,6,6′-tetrasubstituted diphenyl ethers in this way. Even in simple substituted substrates, only trace amounts of the desired products are ever obtained using these synthetic strategies (Scheme 1.4) [57]. The difficulty associated with the construction of the electron-rich 2,2′,6,6′-tetrasubstituted diphenyl ether in **16** is therefore one of the main reasons why **16** has not yet been synthesized.

1.8.2 Synthesis of Benzaldehydes by the Direct Formylation of the Corresponding Aromatic Compounds

The direct formylation of aromatic compounds by electrophilic aromatic substitution is a step-economical reaction for the formation of C–C bonds from C–H bonds in one step. The resulting formyl group can be used as a C1 unit to provide access to a diverse range of interesting products **35**–**39** (Fig. 1.15). Benzaldehydes **34** are therefore incredibly useful synthetic intermediates.

Numerous methods have been developed for the formylation aromatic compounds. Among them, the Vilsmeier-Haack [58, 59] and Duff [60–63] reactions are

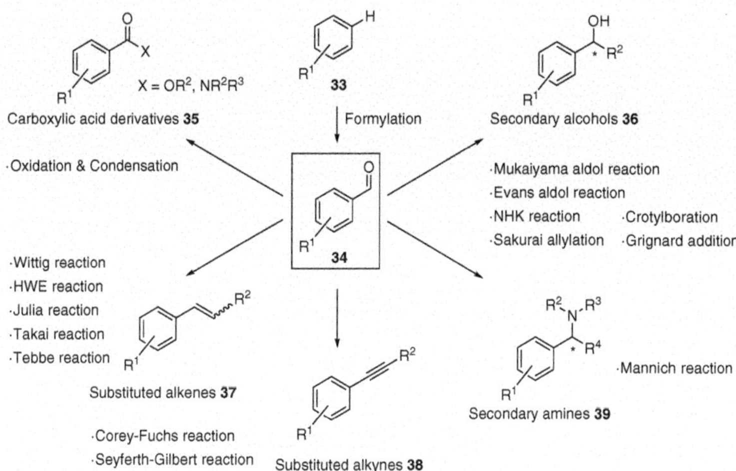

Fig. 1.15 Benzaldehydes as synthetic intermediates

often used for the formylation of electron-rich aromatic compounds. The iminium cations **40** and **41** generated in situ during the Vilsmeier-Haack and Duff reactions, respectively, work as active spices for the formation of intermediate **42**, which undergoes a hydrolysis reaction to afford the corresponding aldehyde **34a** (Scheme 1.5). These reactions are most effective for mono- and disubstituted phenols bearing electron-donating groups because they can be inhibited by the presence of bulky functional groups and electron-withdrawing groups.

The Rieche reaction is a useful method for the formylation of weakly reactive alkoxybenzenes [64, 65]. Active species **43**, which is generated in situ from $Cl_2CHOMe/TiCl_4$, can be captured by a wide range of substrates even at low temperature (Scheme 1.6). Aryl lithium spices **47**, which can be prepared from **33c** using the alkoxyl group as a directing group, can also be used as a good scavenger for formylating reagents (Scheme 1.7). Although formylation reactions of this type can proceed at sterically hindered positions, the substrates themselves are limited to strongly acidic or basic conditions. Notably, the introduced formyl group can work

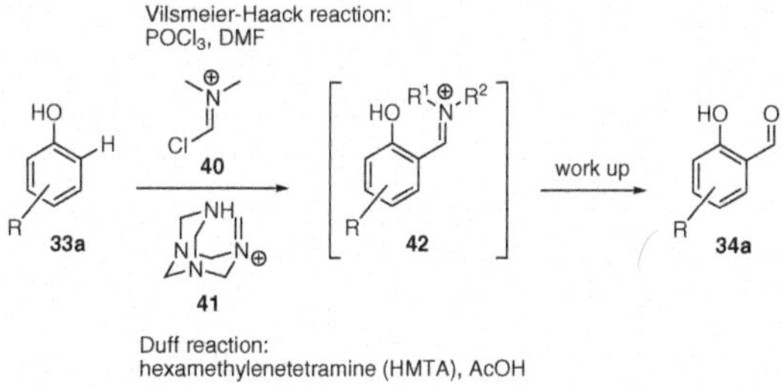

Vilsmeier-Haack reaction: $POCl_3$, DMF

33a 40 41 42 work up 34a

Duff reaction: hexamethylenetetramine (HMTA), AcOH

Scheme 1.5 Synthesis of benzaldehydes via iminium cation intermediates

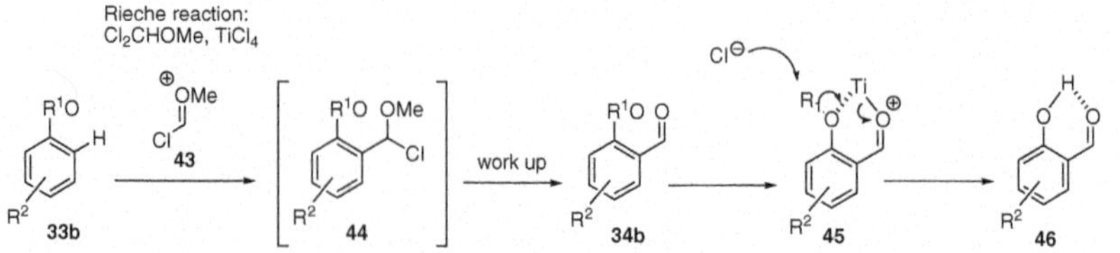

Scheme 1.6 Synthesis of benzaldehydes using a Rieche reaction

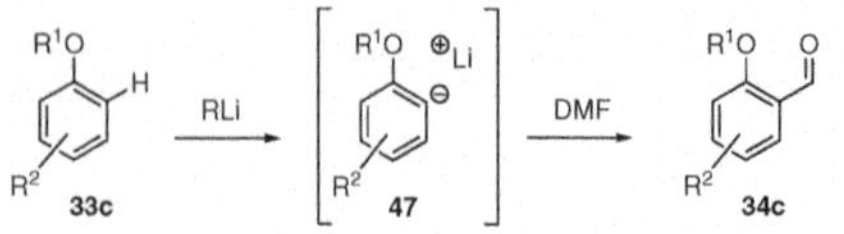

Scheme 1.7 Synthesis of benzaldehydes using aryl lithium species

as a directing group to remove alkyl groups under the acidic conditions used in the Rieche reaction. Although numerous methods are available for the formylation of aromatic compounds, the development of a new method with high reactivity and good functional group tolerance is still required.

1.9 Evaluation of Small Compound/Protein Complexes Using NMR Experiments

1.9.1 NMR Chemical Shift Perturbation Experiments

Recent developments in NMR measurement techniques have enabled researchers to analyze proteins by sequential assignment, and NMR has been used for identifying binding sites in small-molecule ligand/protein and protein/protein complexes. Notably, this technique is conducted in solution, which represents a significant advantage over X-ray crystallography, where the cocrystallization of a ligand in the binding site of a protein is required to develop an understanding of the way in which the ligand is bound to the protein.

NMR chemical shift perturbation [66] is often performed as the analysis of first choice because it only requires samples of the ligand and protein of interest and an NMR spectrometer. The magnetic environments at the interface of an interaction between a ligand and protein can change significantly before and after the formation of the ligand/protein complex. These changes generally manifest themselves as changes in the chemical shifts of the NMR signals belonging to specific residues on the protein, and it is therefore possible to identify the residues at the binding site by comparing the chemical shifts of these residues in the presence and absence of the paired ligand. However, structural changes in the target protein can occur during the formation of the ligand/protein complex, which can lead to changes in the chemical shifts of residues that are positioned far away from the binding site. Furthermore, structural changes can cancel out changes in the chemical shifts of residues located close to the interface of the interaction between the ligand and the protein, and residues involved in the interaction could therefore be missed. With this in mind, even though this experiment is easy to conduct, the resulting information should be treated with caution and used in conjunction with additional data where possible [67].

1.9.2 Spin-Labeling Method

Stable nitroxyl radicals such as TEMPO are not only useful as catalytic oxidants and radical scavengers in organic chemistry but can also be used as tools for tracing specific phenomena spectroscopically. Given that atoms in close proximity to the paramagnetic radicals (i.e., 15–25 Å) are promoted to mitigate their nuclear spins, the intensities of their NMR signals can be decreased significantly [68]. The observation of paramagnetic relaxation enhancement (PRE) effects, which is otherwise known as the spin-labeling method, can therefore be used for extracting long-distance information from small-molecule ligand/protein and protein/protein complexes [69]. By comparing the intensities of the NMR signals in a spin-labeled probe/target complex both before and after the quenching of the radical, it is possible to determine the location of the spin-labeled probe. Furthermore, because this method uses the resulting complex as a standard, it is much less susceptible to structural changes. Spin-labeling reagents **48** and **49** are commercially available and can also be readily synthesized from simple starting materials (Fig. 1.16). Proteins can be easily spin-labeled at specific amino acid residues using click-chemistry or via the formation of a disulfide bond in water. In contrast, molecular probes bearing complicated ligands can be quite difficult to access using similar strategies because of limitations associated with the proposed site of introduction and difficulties with the synthetic tractability. Although important information pertaining to the binding sites at the wide interfaces of PPIs could be directly detected using small-molecule probes of this type, no examples have been reported to date.

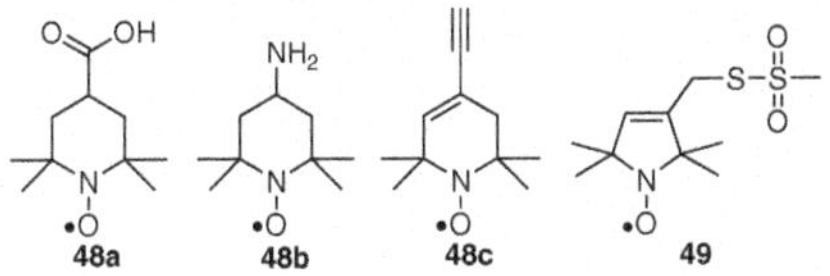

48a 48b 48c 49

Fig. 1.16 Spin-labeling reagents **48** and **49**

1.10 Outline of This Book

As mentioned above, thielocin B1 (**16**) was identified as an inhibitor of PAC3 homodimer. Preliminary in silico studies indicated that **16** specifically interacts with a hill-like structure on the surface of PAC3 homodimer. It is expected that the elucidation of the rare mode of the interaction between thielocin B1 and PAC3 homodimer will provide useful information pertaining to the mechanisms available to PPI inhibitors, and highlight the differences between conventional drug discovery processes targeting well-defined binding pockets on the surfaces of proteins and PPI inhibitors targeting the surfaces themselves. To investigate the validity of the binding model proposed above, it will be necessary to develop a synthetic strategy for the construction of **16** and its derivatives for NMR experiments and further in silico studies.

This chapter has provided a detailed description of the potential of PPIs as drug discovery targets and the difficulties associated with targeting the interactions at the interface of PPIs. Natural products cover a wide area of chemical space, and can therefore be used as valuable tool compounds to provide greater insights into PPIs; not only in terms of their mechanism of interaction but also in terms of identifying potential mechanisms of inhibition. In this way, the HTS of a natural product library led to the identification of compound **16** as an inhibitor of PAC3 homodimer. Furthermore, the suggestion that the mode of interaction of the **16**/PAC3 complex might be different from that of conventional PPI inhibitors further highlights the significance of this study.

Chapter 2 describes the development of our synthetic strategy for the total synthesis of **16**, which was required to supply enough material to complete the NMR studies. The results of NMR chemical shift perturbation experiments involving synthetic **16** are also discussed in this chapter, together with the results of in silico studies based on these NMR data, which allowed for the interaction model of the **16**/PAC3 complex to be further refined.

Chapter 3 provides a detailed account of the synthesis of spin-labeled **16** and the use of this compound to provide further information concerning the binding site in the **16**/PAC3 complex. The results of several NMR experiments involving spin-labeled **16** are also described in this chapter, together with the results of in silico studies based on these data, which provided greater insights into the interaction model of the **16**/PAC3 complex.

Chapter 4 provides a comprehensive evaluation of the efficiency of our newly developed method for the formylation of sterically hindered aromatic substrates, in terms of its scope and limitations towards a variety of substrates.

Chapter 5 provides a summary of all of the areas covered in this book.

References

1. Holton RA, Somoza C, Kim HB, Liang F, Biediger RJ, Boatman PD, Shindo M, Smith CC, Kim S. J Am Chem Soc. 1994;116:1597–8.
2. Holton RA, Kim HB, Somoza C, Liang F, Biediger RJ, Boatman PD, Shindo M, Smith CC, Kim S. J Am Chem Soc. 1994;116:1599–600.
3. Nicolaou KC, Yang Z, Ueno H, Nantermet PG, Guy RK, Claiborone CF, Renaud J, Couladouros EA, Paulvannan K, Sorensen EJ. Nature. 1994;367:630–4.
4. Masters JJ, Link JT, Synder LB, Young WB, Danishefsky SJ. Angew Chem Int Ed. 1995;34:1723–6.
5. Wender PA, Badham NF, Conway SP, Floreancig PE, Glass ET, Gränicher C, Houze JB, Jänichen J, Lee D, Marquess DG, McGrane PL, Meng W, Mucciaro TP, Mühlebach M, Natchus MG, Paulsen H, Rawlims DB, Satkofsky J, Shuker AJ, Sutton JC, Taylor RE, Tomooka K. J Am Chem Soc. 1997;119:2755–6.
6. Wender PA, Badham NF, Conway SP, Floreancig PE, Glass ET, Houze JB, Krauss NE, Lee D, Marquess DG, McGrane PL, Meng W, Natchus MG, Shuker AJ, Sutton JC, Taylor RE. J Am Chem Soc. 1997;119:2757–8.
7. Shiina I, Iwadare H, Sakoh H, Hasegawa M, Tani Y, Mukaiyama T. Chem Lett. 1998;27:1–2.
8. Shiina I, Saitoh K, Fréchard-Ortuno I, Mukaiyama T. Chem Lett. 1998;27:3–4.
9. Morihara K, Hara R, Kawahara S, Nishimori T, Nakamura N, Kusama H, Kuwajima I. J Am Chem Soc. 1998;120:12980–1.
10. Doi T, Fuse S, Miyamoto S, Nakai K, Sasuga D, Takahashi T. Chem Asian J. 2006;1:370–83.
11. Holton RA, Biedoger RJ, Boatman PD. Taxol Sci Appl. 1995; 97–121.
12. Hirata Y, Uemura D. Pure Appl Chem. 1986;58:701–10.
13. Yu MJ, Zheng W, Seletsky BM. Nat Prod Rep. 2013;30:1158–64.
14. Braun P, Gingras A-C. Proteomics. 2012;12:1478–98.
15. Harris SL, Levine AJ. Oncogene. 2005;24:2899–908.
16. Youle RJ, Strasser A. Nat Rev Mol Cell Biol. 2008;9:47–59.
17. Whitty A, Kumaravel G. Nat Chem Biol. 2006;2:112–8.
18. Conte LL, Chothia C, Janin J. J Mol Biol. 1999;285:2177–98.
19. Smith RD, Hu L, Falkner JA, Benson ML, Nerothin JP, Carlson HA. J Mol Graph Model. 2006;24:414–25.
20. Wells JA, McClendon CL. Nature. 2007;450:112–8.
21. Zinzalla G, Thurston DE. Future. Med Chem. 2009;1:65–93.
22. Arkin MR, Wells JA. Nat Rev Drug Dis. 2012;11:173–5.
23. Vu B, Wovkulich P, Pizzolato G, Lovey A, Ding Q, Jiang N, Liu J-J, Zhao C, Glenn K, Wen Y, Tovar C, Packman K, Vassilev L, Graves B. ACS Med Chem Lett. 2013;4:466–9.
24. Olterdolf T, Elmore SW, Shoemaker AR, Armstrong RC, Augeri D, Belli BA, Bruncko M, Deckwerth TL, Dinges J, Hajduk PJ, Joseph MK, Kitada S, Korsmeyer SJ, Kunzer AR, Letao A, Li C, Mitte MJ, Nettesheim DG, Ng S, Nimmer PM, O'Connor JM, Oleksijew A, Petros AM, Reed CR, Shen W, Tahir SK, Thompson CB, Tomaselli KJ, Wang B, Wendt MD, Zhang H, Fesik SW, Rosenberg SH. Nature. 2005;435:677–81.
25. Tse C, Shoemaker AR, Adickes J, Anderson MG, Chen J, Jin S, Johnson EF, Marsh KC, Mitten MJ, Nimmer P, Roberts L, Tahir SK, Xiao Y, Yang X, Zhang H, Fesik S, Rosenberg SH, Elmore SW. Cancer Res. 2008;68:3421–8.
26. Souers AJ, Leverson JD, Boghaert ER, Ackler SL, Catron ND, Chen J, Dayton BD, Ding H, Enschede SE, Fairbrother WJ, Huamg DCS, Hymowitz SG, Jin S, Khaw SL, Kovar PJ, Lam LT, Lee J, Maecker HL, Marsh KC, Mason KD, Mitten MJ, Nimmer PM, Oleksijew A, Park CH, Park C-M, Phillips DC, Roberts AD, Sampath D, Seymour JF, Smith ML, Sullivan GM, Tahir SK, Tse C, Wendt MD, Xiao Y, Xue JC, Zhang H, Humerickhouse RA, Rosenberg SH, Elmore SW. Nat Med. 2013;19:202–8.

27. Hashimoto J, Watanabe T, Seki T, Karasawa S, Izumikawa M, Seki T, Iemura S, Natsume T, Nomura N, Goshima N, Miyawaki A, Takagi M, Shin-ya K. J Biomol Screen. 2009;14:970–9.
28. Ueyama T, Kusakabe T, Karasawa S, Kawasaki T, Shimizu A, Son J, Leto TL, Miyawaki A, Saito N. J Immunol. 2008;181:629–40.
29. Miyawaki A, Karasawa S. Nat Chem Biol. 2007;3:598–601.
30. Izumikawa M, Hashimoto J, Hirokawa T, Sugimoto S, Kato T, Takagi M, Shin-ya K. J Nat Prod. 2010;73:628–31.
31. Hirano Y, Hendil KB, Yashiroda H, Iemura S, Nagane R, Hioki Y, Natsume T, Tanaka K, Murata S. Nature. 2005;437:1381–5.
32. Hirano Y, Hayashi H, Iemura S, Hendil KB, Niwa S, Kishimoto T, Kasahara M, Natsume T, Tanaka K, Murata S. Mol Cell. 2006;24:977–84.
33. Le Tellec B, Battault MB, Courbeyrette R, Guerois R, Marsolier-Kergoat MC, Peyroche A. Mol Cell. 2007;27:660–74.
34. Kusmierczyk AR, Kunjappu MJ, Funakoshi M, Hochstrasser M. Nat Struct Mol Biol. 2008;15:237–44.
35. Romas PC, Dohmen RJ. Structure. 2008;16:1296–304.
36. Yashiroda H, Mizushima T, Okamoto K, Kameyama T, Hayashi H, Kishimoto T, Niwa S, Kasahara M, Kurimoto E, Sakata E, Takagi K, Suzuki A, Hirano Y, Murata S, Kato K, Yamane T, Tanaka K. Nat Struct Mol Biol. 2008;15:228–36.
37. Bonvini P, Zorzi E, Basso G, Rosolen A. Leukemia. 2007;24:838–42.
38. Matsumoto K, Tanaka K, Matsutani S, Sakazaki R, Hinoo H, Uotani N, Tanimoto T, Kawamura Y, Nakamoto S. Yoshida. T J Antibiot. 1995;48:106–12.
39. Genisson Y, Tyler PC, Young RN. J Am Chem Soc. 1994;116:759–60.
40. Genisson Y, Tyler PC, Ball RG, Young RN. J Am Chem Soc. 2001;123:11381–7.
41. Genisson Y, Young RN. Tetrahedron Lett. 1994;35:7747–50.
42. Teshirogi I, Matsutani S, Shirahase K, Fujii Y, Yoshida T, Tanaka K, Ohtani M. J Med Chem. 1996;39:5183–91.
43. Friensner RA, Banks JL, Murphy RB, Halgrem TA, Klicic JJ, Mainz DT, Repasky MP, Knoll EH, Shelley M, Perry JK, Shaw DE, Francis P, Shenkin PS. J Med Chem. 2004;47:1739–49.
44. The results of preliminary simulation are also described in our published paper.
45. Ullmann F. Chem Ber. 1904;37:853–4.
46. Ullmann F, Bielecki J. Chem Ber. 1901;34:2174–85.
47. Ullmann F. Chem Ber. 1903;36:2382–4.
48. Ullmann F, Sponagel P. Chem Ber. 1905;38:2211–2.
49. Bäckvall J-E, Björkman EE, Pettersson L, Siegbahn P. J Am Chem Soc. 1984;106:4369–73.
50. Aranyos A, Old DW, Kiyomori A, Wolfe JP, Sadighi JP, Buchwald SL. J Am Chem Soc. 1999;121:4369–78.
51. Mann G, Incarvito C, Rheingold AL, Hartwig JF. J Am Chem Soc. 1999;121:3224–5.
52. Cristau H-J, Cellier PP, Hamada S, Spindler J-F, Taillefer M. Org Lett. 2004;6:913–6.
53. Hartwig JF. Angew Chem Int Ed. 1998;37:2046–67.
54. Ley SV, Thomas AW. Angew Chem Int Ed. 2003;42:5400–49.
55. Frlan R, Kikelj D. Synthesis. 2006;14:2271–85.
56. Beston MS, Clayden J, Worrall CP, Peace S. Angew Chem Int Ed. 2006;45:5803–7.
57. Sperotto E, Klink GPM, Vries JG, Koten G. Tetrahedron. 2010;66:9009–20.
58. Vilsmeier A, Haack A. Ber Deutsch Chem Ges. 1927;60:119–22.
59. Downie IM, Earle MJ, Heaney H, Shunhaibar KF. Tetrahedron. 1993;49:4015–34.
60. Duff JC, Bills EJ. J Chem Soc. 1932; 1987–88.
61. Duff JC, Bills EJ. J Chem Soc. 1934; 1305–08.
62. Duff JC, Bills EJ. J Chem Soc. 1941; 547–550.
63. Smith WE. J Org Chem. 1972;37:3972–3.

64. Rieche A, Gross H, Höft E. Chem Ber. 1960;93:88–94.
65. Gross H, Rieche A, Mattey G. Chem Ber. 1963;96:308–19.
66. Foster MP, Wuttke DS, Clemens KR, Jahnke W, Radhakrishnan I, Tennant L, Reymond M, Chung J, Wright PE. J Biomol NMR. 1998;12:51–71.
67. Takahashi H, Nakanishi T, Kami K, Arata Y, Shimada I. Nat Struct Biol. 2000;7:220–3.
68. Battiste JL, Wangner G. Biochemistry. 2000;39:5355–65.
69. Jahnke W. ChemBioChem. 2002;3:167–73.

Chapter 2
Total Synthesis of Thielocin B1 and NMR Chemical Shift Perturbation Experiments with PAC3 Homodimer

2.1 Introduction

Thielocin B1 (**1**) was isolated from the fermentation broth of *Thielavia terricola* RF-143 by Yoshida et al. [1] in 1995. The high-throughput screening of a natural products library was performed by Shin-ya et al. [2] in 2009 to identify protein–protein interaction (PPI) inhibitors, and it was revealed that **1** strongly inhibited the formation of the proteasome-assembling chaperon 3 (PAC3) homodimer [3], which is a component of the α-rings of the 20S proteasome [4]. Remarkably, thielocin A1β (**2**) and B3 (**3**) did not exhibit any inhibitory activity towards the same PPI (Fig. 2.1), which indicated that the central core in **1** is in some way critical to the expression of its activity, because the wings to the left and right of all three of these compounds are identical. Additionally, the results of glide docking experiments indicated that **1** recognized specific β-sheet structures on the interface of PAC3, although no cavities are visible in this particular area of the protein. It was envisaged that an NMR chemical shift perturbation experiment with the **1**/PAC3 complex would allow for the characterization of the rare and interesting inhibitory mechanism of **1**. However, it would be necessary to achieve the organic synthesis of **1** to conduct this experiment because the supply of this compound from natural sources is limited. Herein, we report the first total synthesis of **1** and the characterization of its inhibitory mechanism based on the results of NMR chemical shift perturbation experiments involving PAC3 homodimer with synthesized **1** and an in silico docking study.

2.2 Strategy

Thielocin B1 (**1**) is composed of five fully substituted benzene rings, which are connected to each other through one ether and three ester linkages. It is known that highly *ortho*-substituted diphenyl ether and phenyl benzoate compounds are

K. Ohsawa, *Total Synthesis of Thielocin B1 as a Protein–Protein Interaction Inhibitor of PAC3 Homodimer*, Springer Theses, DOI 10.1007/978-4-431-55447-9_2

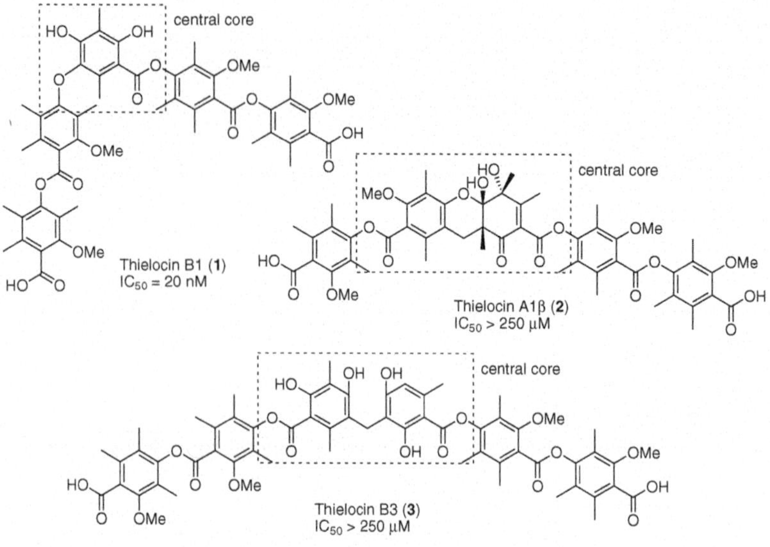

Fig. 2.1 IC_{50} values of several thielocins towards PAC3 homodimer

difficult to form because of steric hindrance. In particular, there have been very few reports pertaining to the synthesis of electron-rich 2,2′,6,6′-tetrasubstituted diphenyl ether moieties using transition metal-catalyzed intermolecular coupling reactions (e.g., Pd or Cu) [5]. It was envisaged, however, that the use of a depsidone skeleton, which is found in a number of natural products, could provide a workable solution to this critical problem. Depsidone **4** possesses a diphenyl ether moiety as part of a seven-membered lactone ring, and the reductive ring opening of the lactone in **4** could therefore provide access to the desired diphenyl ether **5** (Fig. 2.2).

The retrosynthetic strategy for the construction of **1** is shown in Scheme 2.1. Compound **1** was disconnected in three units, including a central core **6**, right wing **7** and left wing **8**. To prevent the nonselective elongation of the side wings, one of

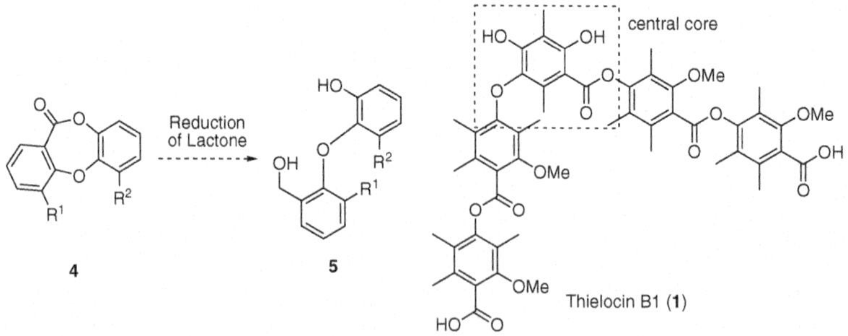

Fig. 2.2 Synthetic strategy for the construction of the 2,2′,6,6′-tetrasubstituted diphenyl ether moiety

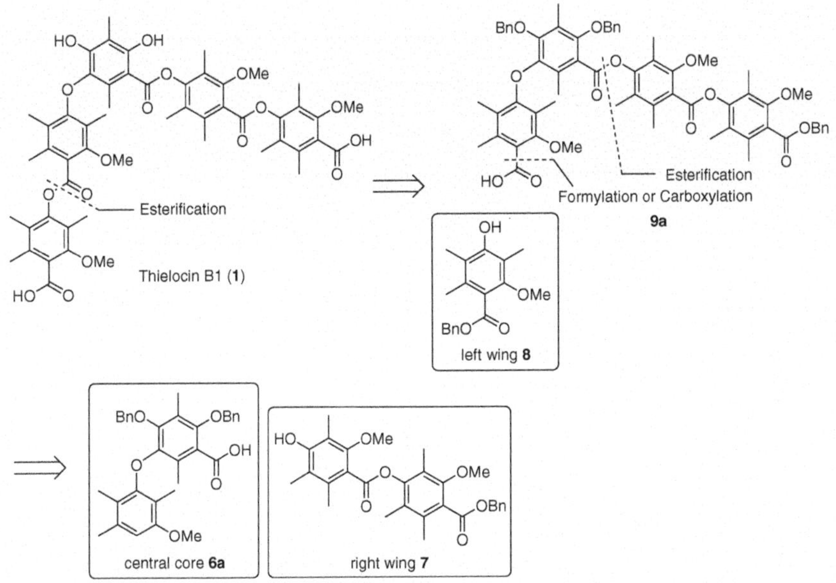

Scheme 2.1 Retrosynthesis of thielocin B1 (**1**)

the carboxyl groups would have to be introduced after the esterification of acid **6a** with phenol **7**. However, highly reactive formylation or carboxylation methods capable of introducing carbonyl-containing functionalities at sterically hindered positions would be required to access **9a**. Given that **1** was expected to be a highly polar compound, benzyl groups were chosen as the protecting groups for the two hydroxyl and two carboxyl groups in **1** under the assumption that they could be readily removed by hydrogenolysis under neutral conditions in the final step of the synthesis.

The synthetic strategies for the construction of the different components of **1** are described below. As mentioned above, **6a** could be synthesized by the chemoselective reduction of the lactone moiety in **10a** without losing the methoxycarbonyl group. Following on from the Friedel-Crafts acylation of **12** with acid **13**, it was envisaged that the resulting ketone **11** could be subjected to an oxidative lactonization according to the procedure reported by Sargent et al. [6] to afford **10a** (Scheme 2.2). For the right wing **7**, which is equivalent to two molecules of the left wing **8**, it was envisaged that the esterification of acid **8** with phenol **14** would give **7**. During the synthesis of these different components, aldehyde **15** was identified as a key synthetic intermediate, because it was used to access compounds **8**, **13** and **14** (Scheme 2.3).

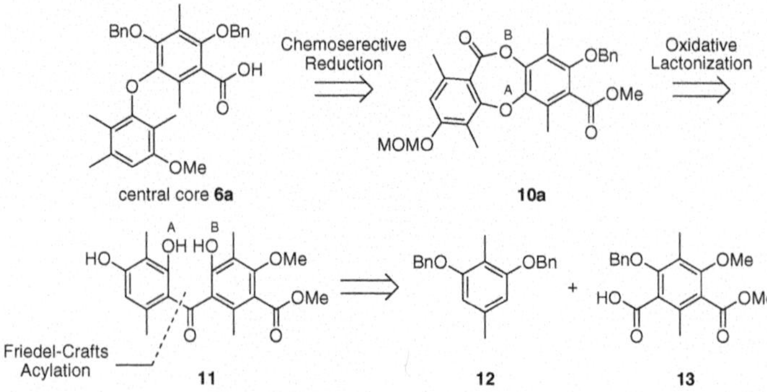

Scheme 2.2 Retrosynthesis of central core **6a**

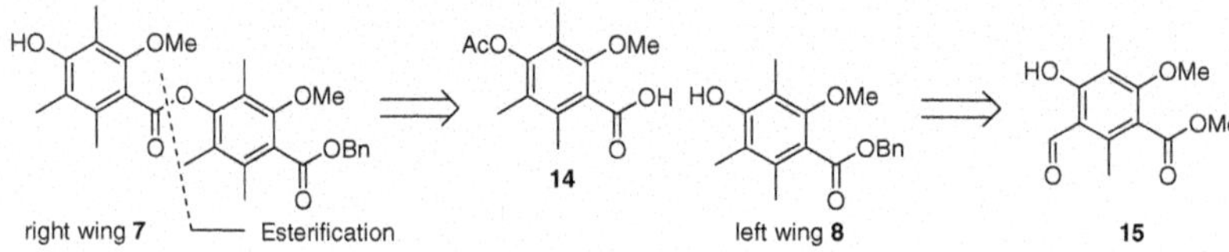

Scheme 2.3 Retrosynthesis of side wings **7** and **8**

2.3 Synthesis of the Central Core of Thielocin B1

2.3.1 Construction of the Depsidone Skeleton

To provide access to the desired depsidone, it was necessary to develop scalable synthetic routes to compounds **12** and **13**. The commercially available methyl 2,4-dihydroxy-3,6-dimethylbenzoate (**16**) bearing two hydroxyl groups was selectively benzylated to afford monobenzyl ether **17**. Subsequent methylation of the other hydroxyl group followed by the cleavage of the benzyl group and a modified Duff reaction [7] afforded key intermediate **15** from **17** in 63 % yield. Protection of the free hydroxyl group as a benzyl ether, followed by a Kraus oxidation [8–10], gave the known acid **13** (Scheme 2.4) [6]. It is noteworthy that the original route developed for the synthesis of **12** had to be abandoned because commercially available **20** was too expensive. The hydrolysis of **12** with KOH proceeded with concomitant decarboxylation to give **20** in 91 % yield. Subsequent protection of the two hydroxyl groups in **20** as benzyl ethers provided the desired product **12** in good yield (Scheme 2.5).

The preparation of depsidone **22** is shown in Scheme 2.6. The Friedel-Crafts acylation of **12** with **13** in the presence of $(CF_3CO)_2O$ afforded benzophenone **21**. Following the cleavage of all three of its benzyl groups by hydrogenolysis, the resulting triol **11** was oxidized with alkaline ferricyanide to give the known lactone

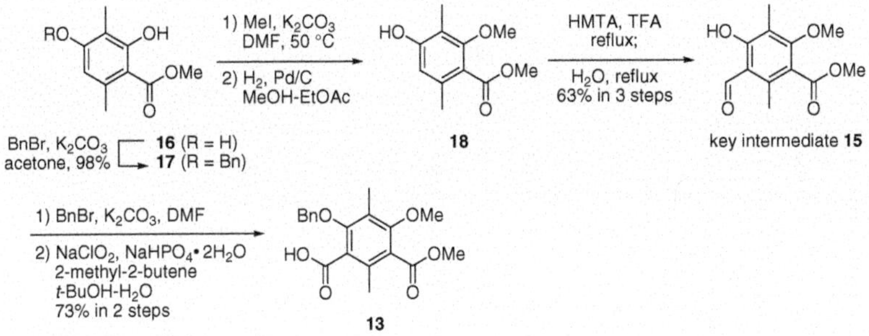

Scheme 2.4 Synthesis of acid **13**

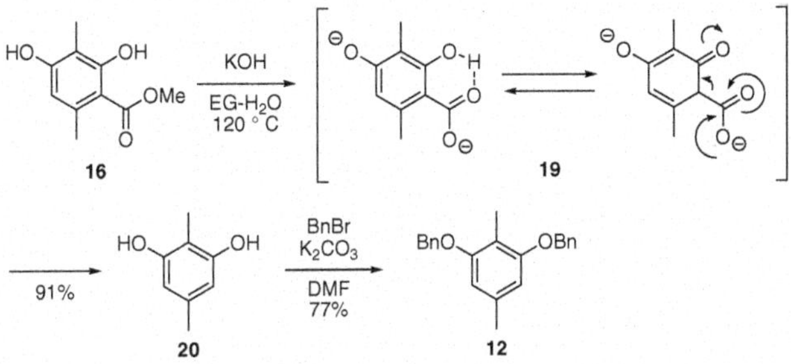

Scheme 2.5 Synthesis of benzyl ether **12**

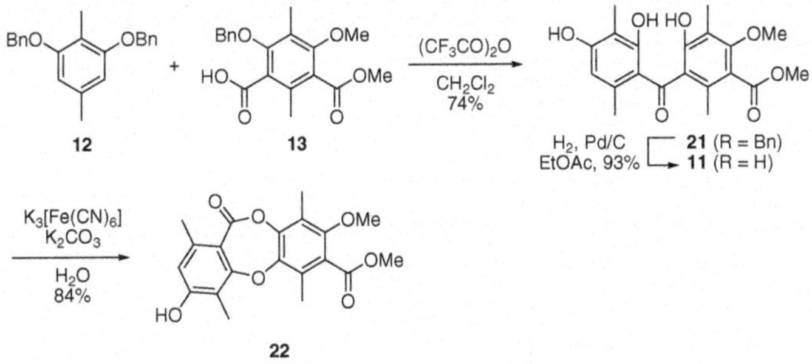

Scheme 2.6 Synthesis of depsidone **22**

22 [6]. A plausible mechanism for this transformation is shown in Scheme 2.7, and is similar to the mechanism suggested by Sargent et al. [11]. Briefly, an initial single electron oxidation would proceed at the more electron-rich benzene ring A. The homolytic formation of a C–O bond followed by further oxidation would give **25** as an intermediate [12, 13]. Subsequent heterolytic (route A) or homolytic (route B) cleavage would result in the immediate cyclization of the resulting C–O bond of **26a** or **26b**, respectively, to give **22**.

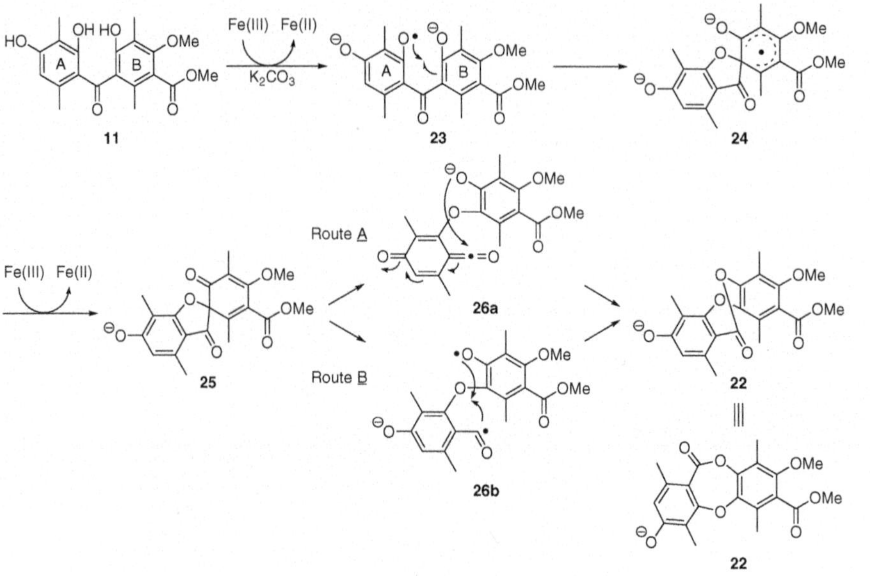

Scheme 2.7 Plausible mechanism for the oxidative lactonization

2.3.2 Construction of Desired Diphenyl Ether Moiety

With the desired depsidone **22** in hand, it was possible to investigate the conversion of this material to the desired biphenyl ether moiety. The method used for the protection of hydroxyl groups in **22** is shown in Scheme 2.8. As shown in Scheme 2.2, it would be preferable to protect hydroxyl group A as a benzyl ether

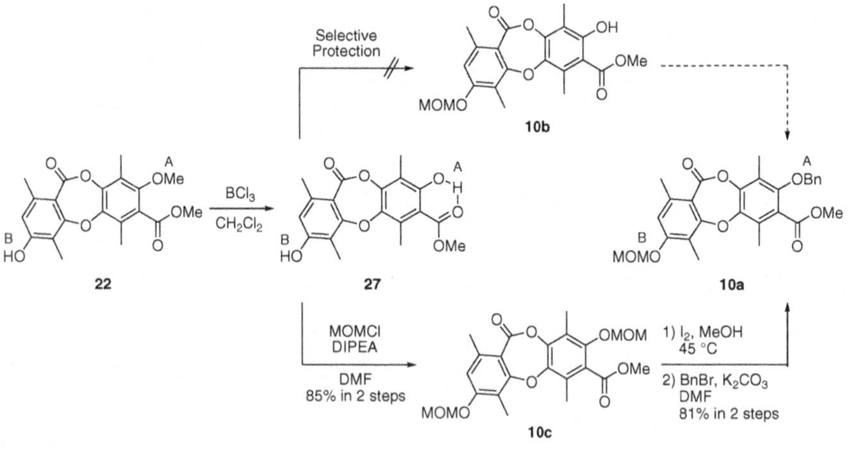

Scheme 2.8 Synthesis of key precursor **10a**

group so that it could be readily removed in the final step. In contrast, hydroxyl group B was protected with a MOM group, which could be easily extricated without losing the benzyl groups at a later stage to allow for the investigation of the formylation or carboxylation strategies. Following the cleavage of the methyl ether in **22** with BCl_3, the selective protection of hydroxyl group B in **27** was examined. Although it was expected that hydroxyl group A would be less reactive than hydroxyl group B because it formed an intramolecular hydrogen bond to the methyl ester, the MOM protection step did not proceed in a sufficiently selective manner to give the desired mono-protected material **10b**. This lack of selectivity was attributed to the high reactivity of MOMCl, and a mixture of two mono-protected compounds was obtained even though the reaction did not reach completion. Based on this lack of selectivity, it was therefore necessary to effect the selective removal of one of the MOM ethers from the bis-MOM ether **10c** using chelating effects. Following a period of extensive investigation, it was found that the use of catalytic HI, which was generated in situ from I_2 in MeOH [14], allowed for the selective cleavage of the MOM ether at hydroxyl group A, which was subsequently benzylated to give key precursor **10a**.

Research into the chemoselective reduction of precursor **10** is summarized in Table 2.1. Given that a phenoxide is a better leaving group than a methoxide, it was

Table 2.1 Chemoselective reduction of lactone **10**

MOMO
A OP
OMe
NaBH$_4$
THF
Conditions
10
HO
A OP
OMe
OH
OMOM
28

Entry	Substrate	P	Temp. (°C)	Time (h)	Product (yield, %)	σ_{para} of OP[a]
1	**10a**	Bn	0	58	**28a** (52), **10a** (26)	−0.42
2	**10a**	Bn	Rt	75	**28a** (<48)[b]	−0.42
3	**10b**	H	Rt	62	Complex mixture	−0.52[c]
4	**10c**	MOM	0	94	**28c** (21), **10c** (33)	–
5	**10d**	Me	Rt	44	**28d** (82)	−0.28
6[d]	**10e**	(4-Cl)Bn	Rt	60	**28e** (<60)[b], **10e** (12)	–
7[e]	**10e**	(4-Cl)Bn	30	103	**28e** (69), **10e** (10)	–

[a] Hammett substituent constant
[b] Inseparable mixture of desired **28** and a small amount of unknown byproduct
[c] Constant of O^-
[d] 38 mM
[e] 10 mM

envisaged that the phenyl ester could be reduced prior to the methyl ester. In an initial attempt to influence the selectivity of the reduction, the mildly reactive reducing agent $NaBH_4$ was used to affect the reduction of **10a** under low-temperature conditions. Although these conditions provided access to the desired **28a** in moderate yield, the reproducibility of the reaction was found to be poor (Table 2.1, entry 1). Furthermore, the reaction gave an inseparable mixture of **28a** and an unknown byproduct when it was conducted at room temperature (Table 2.1, entry 2). Several other reducing agents such as DIBAL-H were also investigated but failed to give pure **28a**. The focus of the investigation then shifted towards the nature of the protecting group for hydroxyl group A. Given that highly electron-donating protecting groups can lead to an increase in the electron density of the benzene ring, the more electron-rich phenol would perform poorly as a leaving group compared with the corresponding system with one less electron-donating ether (Fig. 2.3). To confirm the influence of the protecting groups, the reduction of several other substrates **10** bearing different protecting groups at hydroxyl group A was investigated. Although the application of the reduction conditions to the methyl esters in phenol **10b** and MOM ether **10c** gave a complex mixture or poor selectivity, respectively (Table 2.1, entries 3 and 4), methyl ether **10d** was reduced smoothly to provide the desired compound **28d** exclusively (Table 2.1, entry 5). These results for the different alkoxy groups were also found to be in agreement with their Hammett substituent parameters [15]. Considering these results, the 4-chlorobenzyl group was selected as the best protecting group for hydroxyl group A because it could be readily cleaved at a later stage by hydrogenolysis. Furthermore, the electron-withdrawing effect of the Cl substituent could lead to an improvement in the yield [16]. As expected, 4-chlorobenzyl ether **10e** was successfully converted to **28e** in a yield comparable with that of benzyl ether **10a** (Table 2.1, entry 6 vs. entry 2). Moreover, the formation of the inseparable unknown product was suppressed by running the reaction under dilute conditions (Table 2.1, entry 7). These dilute conditions also appeared to suppress the occurrence of unwanted intermolecular side reactions. Disappointingly, however, compound **28e** was not reproducibly obtained in a scalable reaction, and the reaction had to be carried out over several smaller batches to provide enough material to investigate the subsequent steps.

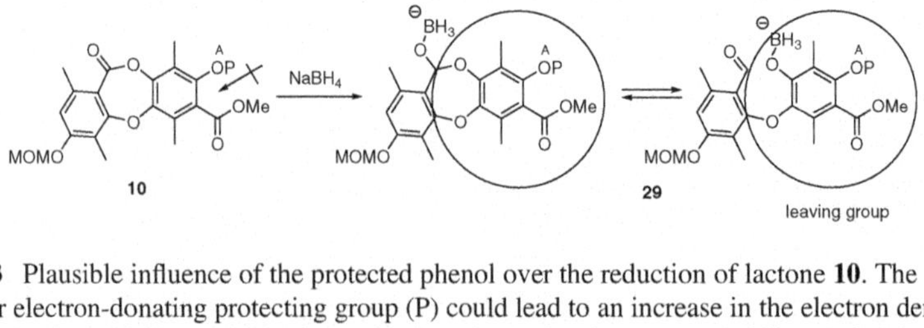

Fig. 2.3 Plausible influence of the protected phenol over the reduction of lactone **10**. The use of a stronger electron-donating protecting group (P) could lead to an increase in the electron density on the phenyl ring, which would lead to a decrease in the leaving group activity on the phenol

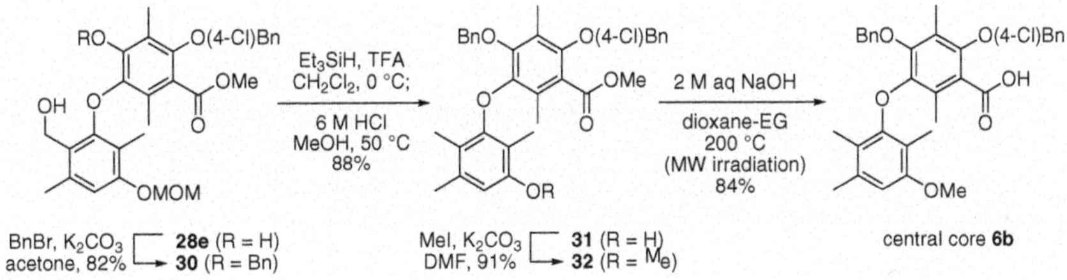

Scheme 2.9 Synthesis of central core **6b**

The central core **6b** was synthesized from **28e**, as shown in Scheme 2.9. The phenolic hydroxyl group in **28e** was selectively benzylated to afford **30**. The reduction of the benzylic alcohol with Et_3SiH [17] and removal of the MOM group under acidic conditions were performed in a one-pot reaction to give **31**. The subsequent methylation of **31** gave the corresponding methyl ether **32**. The basic hydrolysis of the methyl ester in **32** was examined, and the reaction was found to proceed smoothly under microwave irradiation conditions at 200 °C to furnish the central core **6b** in high yield [18], whereas the 2,6-disubstituents of the methyl benzoate effectively prevented the hydrolysis from occurring under conventional conditions [19, 20].

2.3.3 Introduction of the Carboxyl Group for Central Core

As mentioned in Sect. 2.2, the late-stage introduction of one more carboxyl group would be required at the sterically hindered position of the central core to allow it to undergo a condensation reaction with the left wing **8**. To establish a suitable method for this transformation, a preliminary investigation was conducted using compounds **31** and **32**. Considering the reactivity of the electrophile, it was envisaged that a two-step synthesis would be required for this transformation involving sequential formylation and Kraus oxidation reactions.

Several conditions were investigated for the formylation of **31** and **32**, and the results are summarized in Table 2.2. Conventional formylation methods such as the Vilsmeier-Haack reaction [21], Reimer-Tiemann reaction [22–25] and a modified version of the Duff reaction failed to provide any of the desired aldehyde product from **31**, which is the least sterically hindered of the two compounds (Table 2.2, entries 1–3). In contrast, Rieche conditions using $Cl_2CHOMe–TiCl_4$ [26, 27] provided aldehyde **35** in low yield (Table 2.2, entry 4). However, the use of these strongly acidic conditions also led to the debenzylation of the product to give **35**. The use of milder Lewis acid such as $BF_3{\cdot}OEt_2$ was investigated, but these acids could not sufficiently activate Cl_2CHOMe, and **31** was recovered quantitatively from these reactions (Table 2.2, entry 5). Given that the possibility of competition between *O*- and *C*-formylation could potentially complicate the reaction system, the

Table 2.2 Formylation of central cores **31** and **32**

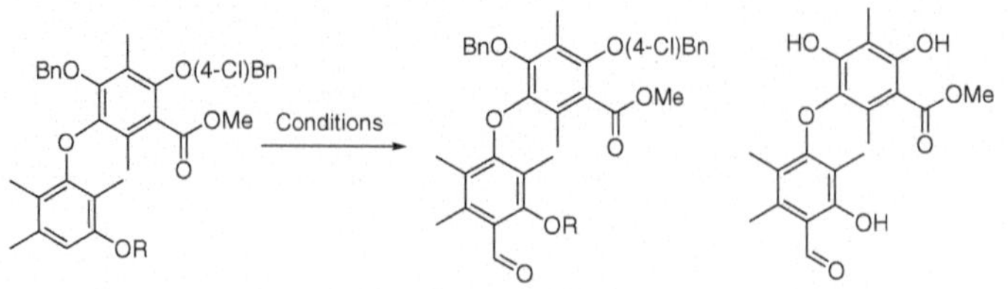

31 (R = H)
32 (R = Me)
33 (R = H)
34 (R = Me)
35

Entry	Substrate	Reagents	Solv.	Temp. (°C)	Time (h)	Product (yield, %)
1	**31**	$POCl_3$	DMF	80	7.5	N.R.
2	**31**	NaOH	$CHCl_3$–H_2O	60	24	N.R.
3	**31**	HMTA	TFA	50	1.5	Decomp.
4	**31**	Cl_2CHOMe $TiCl_4$	CH_2Cl_2	Rt	8	**35** (9)
5	**32**	Cl_2CHOMe $BF_3{\cdot}OEt_2$	CH_2Cl_2	Rt	22	N.R.
6	**32**	Cl_2CHOMe $SnCl_4$	CH_2Cl_2	0	1	**34** (21)
7	**32**	Cl_2CHOMe AgOTf	CH_2Cl_2	−20	4	**34** (45)
8	**32**	Cl_2CHOMe AgCl	CH_2Cl_2	Rt	7.5	N.R.
9	**32**	Cl_2CHOMe $AgOCOCF_3$	CH_2Cl_2	35	12	N.R.
10	**32**	Cl_2CHOMe $AgNTf_2$	CH_2Cl_2	0	12	Decomp.
11	**32**	Cl_2CHOMe $AgClO_4$	CH_2Cl_2	Rt	4	Decomp.

methyl ether **32** was selected as an alternative model substrate. The formylation of **32** using Cl_2CHOMe–$SnCl_4$ [26, 27] gave the desired aldehyde **34** together with the debenzylated byproducts (Table 2.2, entry 6), which suggested that it would not be possible to avoid the debenzylation of the product using the reported Rieche conditions with a hard Lewis acid. For this reason, the focus of the optimization work shifted towards the use of soft Lewis acids, which would have a lower affinity for the O atoms of the benzyl ethers. Following an extensive period of investigation with Ag salts, it was discovered that **32** could be successfully formylated with Cl_2CHOMe–AgOTf to afford **34** in moderate yield without losing the benzyl groups (Table 2.2, entry 7). Several other Ag salts were also tested, including AgCl,

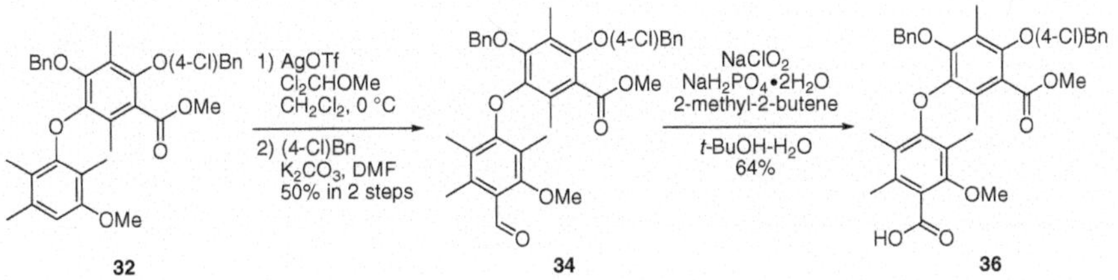

Scheme 2.10 Scalable synthesis of acid **36**

$AgOCOCF_3$, $AgNTf_2$ and $AgClO_4$, but all of these salts failed to provide any of the desired aldehyde (Table 2.2, entries 8–11) [28].

Although a new method had been successfully developed for the synthesis of **34**, the partial removal of the 4-chlorobenzyl group was observed when the reaction was conducted on a larger scale, which was attributed to the TfOH generated during the course of the reaction [29]. Given that the coexistence of bulky Lewis bases such as 2,6-di-*tert*-butylpyridine can deactivate the formylating species, the Kraus oxidation was performed to give acid **36** after the benzylation of the partially deprotected phenol with 4-chlorobenzyl chloride (Scheme 2.10). Several other C–C bond-forming reactions were also investigated for the installation of the carbonyl group, including a haloformylation reaction [30] and the rearrangement of the MOM group [31, 32], but all of these strategies failed to provide a new C–C bond at the sterically hindered position. For this reason, the AgOTf-promoted formylation followed by a Kraus oxidation was deemed to be the best route for the synthesis of **1**.

2.4 Synthesis of the Side Wings

The synthesis of left wing **8**, which is also a component of the right wing **7**, was the first of the two wings to be synthesized. Hydrogenolysis of the common intermediate **15** gave **37**, although it should be mentioned that the reproducibility of this reaction was quite low. Subsequent alkaline hydrolysis under microwave irradiation conditions afforded the corresponding decarboxylated compound **38** as the major product with only a trace amount of desired product **39** being detected. The phenolic anion generated by the treatment of **37** with NaOH could readily pick up a proton on its phenyl ring prior to the hydrolysis reaction because of the steric hindrance of its *ortho*-substituents. Demethylation in a Krapcho fashion (path A) or hydrolysis at the position next to the less sterically hindered sp^3 carbon (path B) could then proceed, and further decarboxylation under heating conditions would provide **38** (Scheme 2.11).

Based on these results, the alkaline hydrolysis of the protected phenol was also examined. Benzyl ether **41** was quantitatively prepared from **15** according to the one-pot procedure for the reduction of salicylaldehyde reported by Kijima et al. [33], followed by benzylation of the resulting phenol **39**. As expected, hydrolysis of

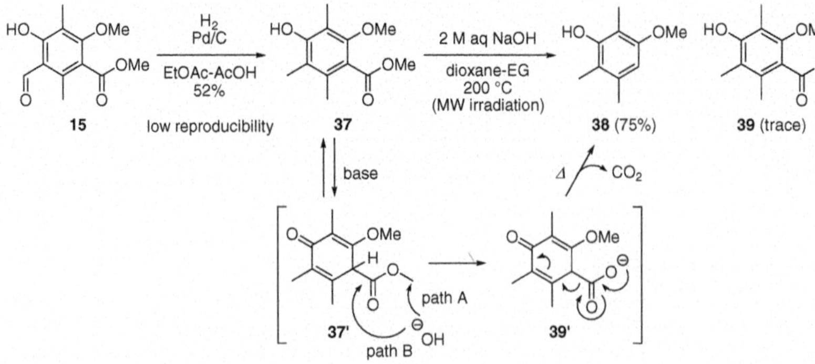

Scheme 2.11 Decarboxylation of **37** under heating conditions

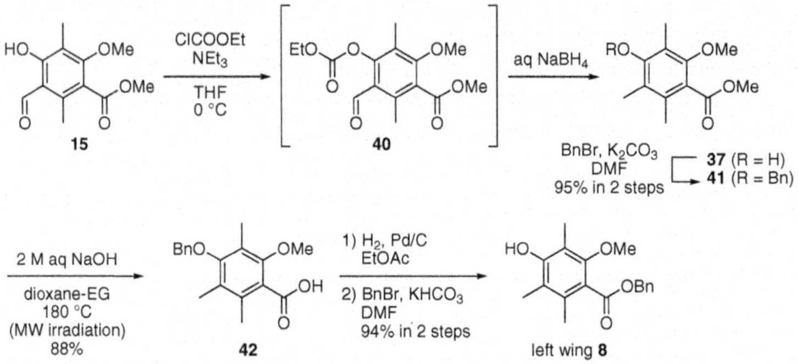

Scheme 2.12 Synthesis of left wing **8**

the methyl ester proceeded without decarboxylation to provide acid **42**. Cleavage of the benzyl group followed by the selective benzylation of the carboxyl group afforded the desired left wing **8** (Scheme 2.12).

The route used for the synthesis of the right wing **7** from synthesized **8** is shown in Scheme 2.13. Compound **8** was converted to acid **14** by the acetylation of the phenol followed by the debenzylation of the benzyl ester. The esterification of acid **14** with phenol **8** proceeded successfully in the presence of the condensation

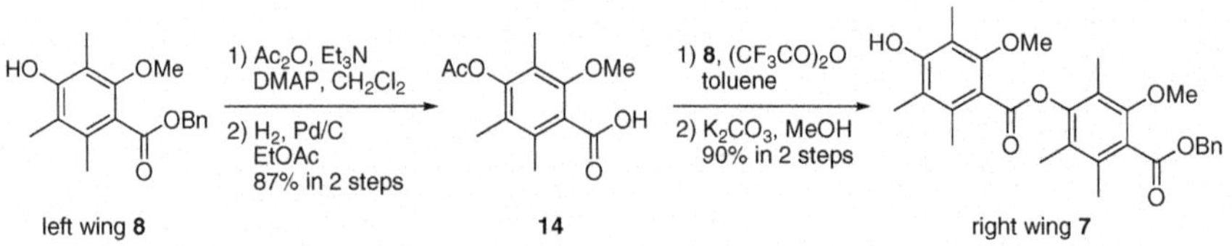

Scheme 2.13 Synthesis of right wing **7**

reagent $(CF_3CO)_2O$ [34] despite the steric hindrance of both coupling partners. Subsequent methanolysis to remove the acetyl group furnished the desired compound **7** in high yield.

2.5 Total Synthesis of Thielocin B1

With all of the components needed for the synthesis of **1** in hand, the focus of the work shifted towards the attachment of the side wings to the central core. Several different conditions were investigated for the esterification of acid **6b** with phenol **7**, and the results are summarized in Table 2.3. Although the use of excess $(CF_3CO)_2O$ at room temperature provided a poor yield of the desired product **43** (Table 2.3, entry 1), increasing the temperature to 80 °C allowed for the near quantitative formation of the desired compound (Table 2.3, entry 2). In contrast, the Corey-Nicolaou reaction [35, 36], which is also conducted under acidic conditions, failed to provide any of the desired product (Table 2.3, entry 3). The mixed anhydride containing a trifluoroacetyl group performed well as a highly active species because of its high electron-withdrawing effect and low steric hindrance.

The total synthesis of thielocin B1 (**1**) is shown in Scheme 2.14. The newly developed formylation method using Cl_2CHOMe-AgOTf was found to be particularly effective for **43**, and subsequent benzylation of the partially formed phenol with 4-chlorobenzyl chloride afforded aldehyde **44** in moderate yield. Subsequent oxidation of **44**, followed by the esterification of the resulting acid **9b** with phenol **8** in the presence of excess $(CF_3CO)_2O$ at 80 °C, provided **45**. Finally, all three of the benzyl groups were removed by hydrogenolysis, thus completing the first total synthesis of thielocin B1 (**1**). A mixture of EtOH and EtOAc was identified as the best solvent for the hydrogenolysis reaction to dissolve both **45** and **1**. The spectral data for synthesized **1** were found to be in good agreement with those of the natural product.

Table 2.3 Esterification of acid **6b** with phenol **7**

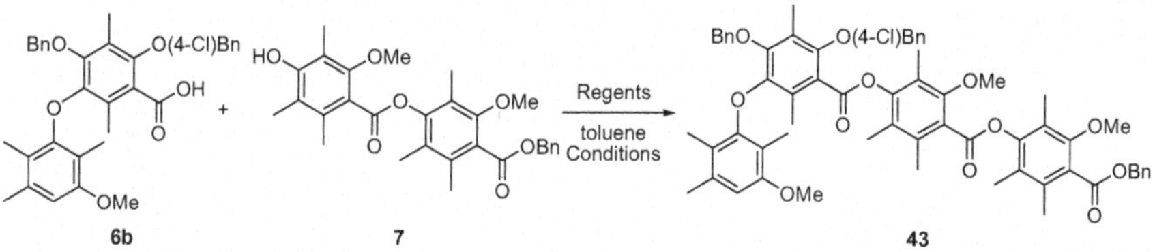

Entry	Reagents	Temp. (°C)	Time (h)	Yield (%)
1	$(CF_3CO)_2O$ (50 equiv)	Rt	29	28
2	$(CF_3CO)_2O$ (40 equiv)	80	9	97
3	$(PyS)_2$, PPh_3, $AgClO_4$	80	11	N.R.

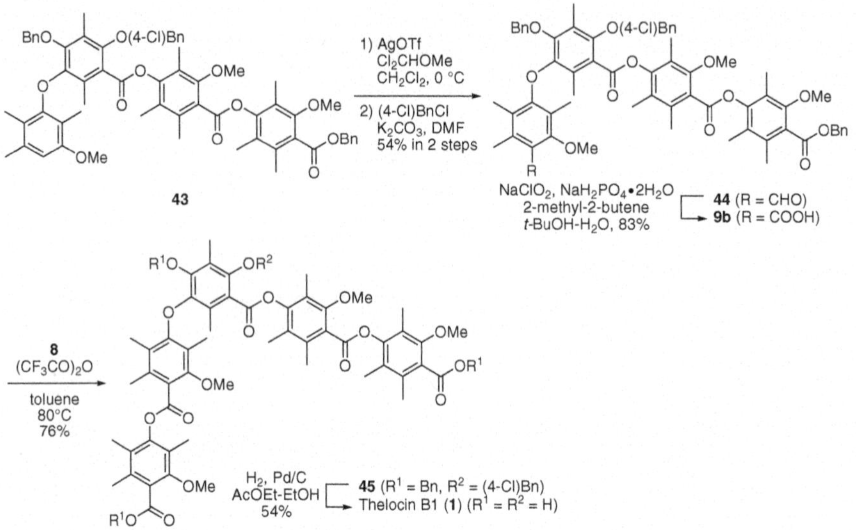

Scheme 2.14 Total synthesis of thielocin B1 (**1**)

The inhibitory activity of synthesized **1** towards PAC3 homodimer was evaluated in vitro using a protein complementation assay with monomeric Kusabira-Green fluorescent protein [2]. The IC_{50} value of the synthesized material was determined to be 40 nM, which was in good agreement with that of natural **1** (20 nM).

2.6 NMR Chemical Shift Perturbation Experiment of PAC3 with Thielocin B1

NMR chemical shift perturbation experiments were conducted with PAC3 homodimer both in the presence and absence of synthesized **1** to elucidate the characteristics of its inhibition mechanism. A solution of **1** in methanol-d_4 (10 mM) was titrated against ^{15}N-labeled PAC3 (0.2 mM for the monomer) to give final concentrations of **1** of 0.1, 0.2, 0.4 and 0.8 mM. These samples were then analyzed by ^{1}H-^{15}N HSQC spectroscopy, which revealed changes in the NMR spectra of PAC3 as the concentration of **1** increased. Further increasing the concentration of **1** resulted in the formation of insoluble material, which was presumably derived from the aggregation of destabilized PAC3, and the ^{1}H–^{15}N HSQC spectra of PAC3 were therefore compared in the presence (0.8 mM) and absence of **1** (Fig. 2.4a). Changes in the chemical shifts were quantified as $[(\Delta\delta_H)^2 + (0.2\Delta\delta_N)^2]^{1/2}$, where δ_H and δ_N are the observed chemical shift changes for ^{1}H and ^{15}N, respectively. As a result, notable changes were observed in the chemical shifts ($[(\Delta\delta_H)^2 + (0.2\Delta\delta_N)^2]^{1/2} > 0.025$ ppm)

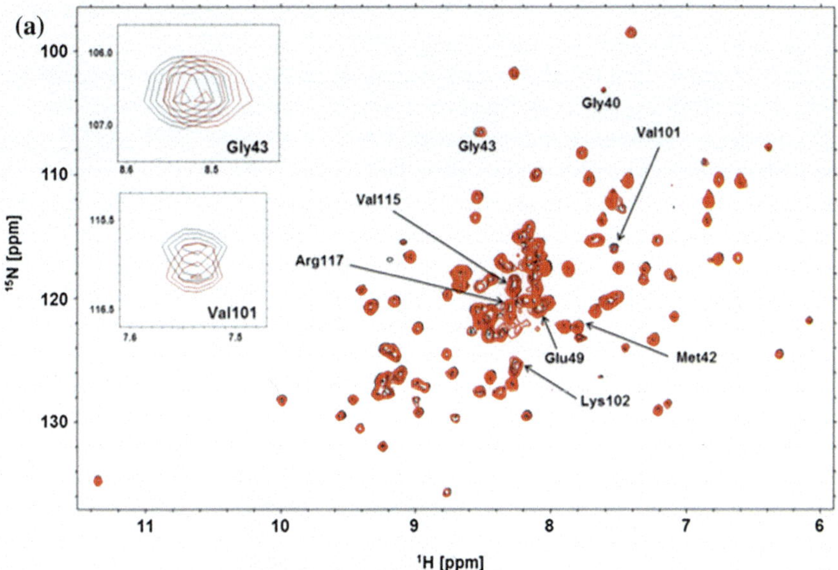

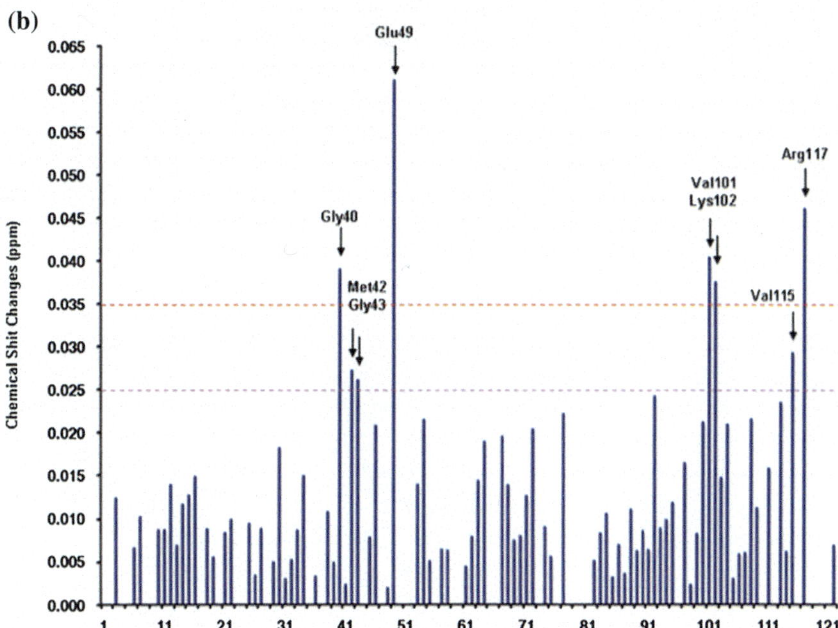

Fig. 2.4 NMR chemical shift perturbation experiment for PAC3 homodimer with **1**. Significant changes in the chemical shifts, i.e., $[(\Delta\delta_{H})^2 + (0.2\Delta\delta_{N})^2]^{1/2} > 0.025$ ppm, have been labeled. **a** ^{1}H–^{15}N NSQC spectra of PAC3 homodimer both in the absence (*black*) and presence (*red*) of four molar equivalents (for the monomer) of **1**. **b** Differences in the chemical shifts according to the equation $[(\Delta\delta_{H})^2 + (0.2\Delta\delta_{N})^2]^{1/2}$

of eight amino acid residues on PAC3, including Gly40, Met42, Gly43, Glu49, Val101, Lys102, Val115 and Arg117 (Fig. 2.4b).

The interaction site of PAC3 with **1** was subsequently examined based on the results of the NMR experiments. However, it was difficult to explain the mechanism of the interaction between **1** and monomeric PAC3 (PDB code: 2Z5E_A) [3] using these data because the eight residues identified by the NMR study appeared to be scattered over the surface of monomeric PAC3 (Fig. 2.5a). In contrast, the mapping of these data onto PAC3 homodimer (PDB code: 2Z5E) [3] indicated that these residues were located in close proximity to each other at the interface of PAC3

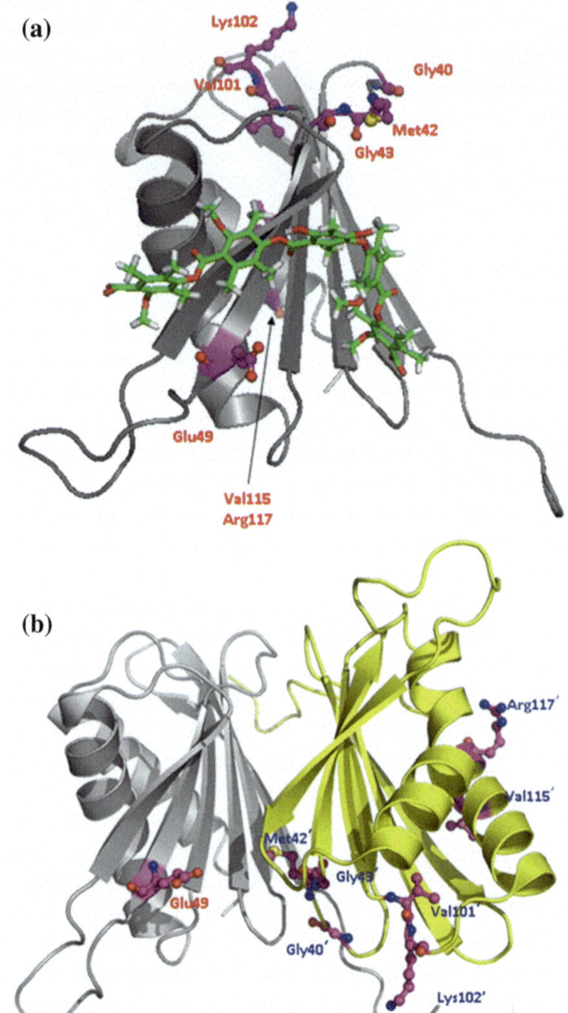

Fig. 2.5 Mapping of the residues exhibiting significant changes in their chemical shifts on PAC3. Gly40, Met42, Gly43, Glu49, Val101, Lys102, Val115 and Arg117 are shown as carbons in magenta (ball and stick models). **a** Mapping of the residues onto monomeric PAC3 (PDB code: 2Z5E_A). **1**, located at the proposed binding site, is shown as carbons in *green* (stick model). **b** Mapping of the residues onto PAC3 homodimer (PDB code: 2Z5E)

homodimer (Fig. 2.5b). These results therefore suggested that the approach of **1** to PAC3 homodimer could be observed using NMR chemical shift perturbation experiments.

2.7 Docking Study for Thielocin B1 and PAC3 Homodimer

Based on the results of the NMR chemical shift perturbation experiments, an in silico docking study was performed for **1** and PAC3 homodimer using the Molecular Operating Environment program [37]. The results of the docking revealed that the most likely pose for **1** involved the molecule spanning the interface of PAC3 homodimer, which suggested that the PPI inhibition mechanism of **1** was as suggested in Fig. 2.6 and in conjunction with the model for the interaction of **1** with monomeric PAC3. In this way, the initial approach of **1** would bring it into close proximity to the Glu49, Gly40′, Met42′ and Gly43′ residues, which are located on the interface of PAC3 homodimer. The insertion of the left wing of **1** into the interface would induce the dissociation of PAC3 homodimer to monomeric PAC3. Changes in the chemical shifts of the Val101′ and Lys 102′ residues would be affected by the movement of the nearby Gly40′, Met42′ and Gly43′ residues. Given that Val115′ and Arg117′ are located at the flexible C-terminal site, they could be susceptible to structural changes on PAC3.

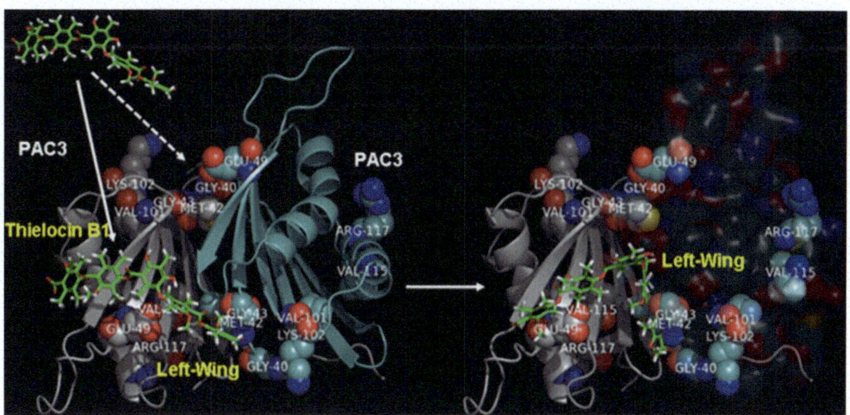

Fig. 2.6 Proposed mechanism of inhibition based on an in silico docking study for **1** (stick model, carbons in *green*)/PAC3. Residues exhibiting significant changes in their chemical shifts are shown as CPK models. The figures to the *left* and *right* show the poses of the **1**/PAC3 homodimer and **1**/monomeric PAC3 complexes, respectively

2.8 Summary

In conclusion, the first total synthesis of thielocin B1 (**1**) has been accomplished. The 2,2′,6,6′-tetrasubstituted diphenyl ether moiety in **1** was synthesized from a depsidone skeleton via the chemoselective reduction of the lactone without losing the methyl ester. It was found that the protecting group of the phenolic hydroxyl group had a significant influence on the reactivity of this reduction. During the development of a synthetic strategy for elongating the side wings, an efficient reaction was discovered for the formylation of sterically hindered aromatic compounds using Cl_2CHOMe–AgOTf. NMR chemical shift perturbation experiments were conducted with ^{15}N-labeled PAC3 homodimer both in the presence and absence of synthesized **1**, and significant changes in the chemical shifts were observed in eight residues at the interface of PAC3 homodimer. Most of these residues were located at the interface of PAC3 homodimer, which indicated that thielocin B1 promotes the dissociation of PAC3 homodimer to monomeric PAC3 following its initial interaction with PAC3 homodimer. However, further studies will be required to validate the proposed mechanism of inhibition.

2.9 Experimental Section

2.9.1 General Techniques

All commercially available reagents were used as received. Dry THF and CH_2Cl_2 (Kanto Chemical Co.) were obtained by passing commercially available pre-dried, oxygen-free formulations.

Microwave irradiation was performed with Biotage Initiator^{TM}.

All reactions were monitored by thin-layer chromatography carried out on 0.2 mm E. Merck silica gel plates (60F-254) with UV light, and visualized by *p*-anisaldehyde H_2SO_4-ethanol solution. Column chromatography and flash column chromatography were carried out with silica gel 60 N (Kanto Chemical Co. 100–210 μm) and silica gel 60 N (Kanto Chemical Co. 40–50 μm), respectively.

^1H NMR spectra (400 MHz and 500 MHz) and ^{13}C NMR spectra (100 MHz and 125 MHz) were recorded on JEOL JNM-AL400 and JEOL ECP-500 spectrometers, respectively, in the indicated solvent. Chemical shifts (δ) are reported in units parts per million (ppm) relative to the signal for internal tetramethylsilane (0.00 ppm for ^1H) for solutions in chloroform-*d*. NMR spectral data are reported as follows: chloroform-*d* (77.0 ppm for ^{13}C), methanol-d_4 (3.30 ppm for ^1H and 49.0 ppm for ^{13}C) when internal standard is not indicated. Multiplicities are reported by the following abbreviations: s (singlet), d (doublet), m (multiplet), brs (broad singlet), *J* (coupling constants in Hertz).

Mass spectra and high-resolution mass spectra were measured on JEOL JMS-DX303 (for EI), MS-AX500 (for FAB), SYNAPT G2 HDMS instruments (for

ESI). IR spectra were recorded on a JASCO FTIR-8400. Only the strongest and/or structurally important absorption are reported as the IR data afforded in cm^{-1}. Melting points were measured on Round Science RFS-10, and are uncorrected.

2.9.2 Experimental Methods

Methyl 4-benzyloxy-2-hydroxy-3,6-dimethylbenzoate (17).
To a solution of methyl 2,4-dihydroxy-3,6-dimethylbenzoate (6.00 g, 30.6 mmol) in acetone (50 mL) were added K_2CO_3 (5.07 g, 36.7 mmol, 1.2 equiv) and BnBr (4.36 mL, 36.7 mmol, 1.2 equiv) at room temperature under an argon atmosphere. After being stirred at the same temperature for 22 h, the reaction mixture was filtered through a pad of Celite®, and the filtrate was concentrated in vacuo. The resulting residue was diluted with EtOAc, and acidified with 3 M aq HCl. The organic layer was separated, and the aqueous layer was extracted with EtOAc twice. The combined organic layers were washed with saturated aq $NaHCO_3$ and brine, dried with $MgSO_4$, and filtered. The filtrate was concentrated in vacuo, and the resulting residue was purified by column chromatography on silica gel (eluted with hexane/EtOAc = 19:1) to afford benzyl ether **17** (8.56 g, 29.8 mmol, 98 %) as a white solid. mp 79–80 °C [lit 79.5–80 °C] [6]; ^{1}H NMR (400 MHz, $CDCl_3$) δ 11.8 (1H, s), 7.31–7.44 (5H, m), 6.35 (1H, s), 5.12 (2H, s), 3.93 (3H, s), 2.51 (3H, s), 2.15 (3H, s); ^{13}C NMR (100 MHz, $CDCl_3$) δ 172.5, 162.3, 160.6, 140.0, 136.9, 128.6, 127.9, 127.0, 111.4, 107.1, 105.7, 69.9, 51.8, 24.6, 8.0; IR (Neat) 3,391, 2,926, 1,651, 1,623, 1,577, 1,453, 1,448, 1,303, 1,278, 1,131, 802 cm^{-1}; HREIMS calcd for $C_{17}H_{18}O_4$ 286.1205, found 286.1197.

Methyl 3-formyl-4-hydroxy-6-methoxy-2,5-dimethylbenzoate (15).
To a solution of phenol **17** (54.0 g, 189 mmol) in DMF (200 mL) were added K_2CO_3 (78.4 g, 567 mmol, 3.0 equiv) and MeI (25.8 mL, 415 mmol, 2.2 equiv) at room temperature under an argon atmosphere. After being stirred at 50 °C for 13 h, the reaction mixture was cooled to room temperature, and filtered through a pad of Celite®. The filtrate was diluted with EtOAc, and acidified with 3 M aq HCl. The organic layer was separated, and the aqueous layer was extracted with EtOAc twice. The combined organic layers were washed with brine twice, saturated aq $NaHCO_3$ and brine, dried with $MgSO_4$, and filtered. The filtrate was concentrated in vacuo to give methyl ether as a yellowish oil. The crude methyl ether was used for next reaction without further purification.

To a solution of crude benzyl ether in EtOAc (40 mL) was added 10 % palladium on carbon (14.0 g, 26 wt%), and poured MeOH (120 mL) slowly over stirring at room temperature. The reaction vessel was purged with hydrogen 7 times. After being stirred at room temperature for 22 h, the reaction mixture was filtered through a pad of Celite®. The filtrate was concentrated in vacuo to give phenol as a white solid. The crude phenol was used for next reaction without further purification.

To a solution of crude phenol in TFA (200 mL) was added HMTA (29.1 g, 208 mmol, 1.1 equiv) at room temperature under an argon atmosphere. After being stirred at reflux for 2 h, the reaction mixture was cooled to room temperature, and concentrated in vacuo. To the resulting residue was poured water (200 mL) at room temperature. After being stirred at reflux for 16 h, the reaction mixture was cooled to room temperature, diluted with EtOAc. The organic layer was separated, and the aqueous layer was extracted with EtOAc twice. The combined organic layers were basified with 2 M aq NaOH. The organic layer was separated, washed with brine, dried with $MgSO_4$, and filtered. The filtrate was concentrated in vacuo, and the resulting residue was recrystallized from CH_2Cl_2/hexane to afford aldehyde **15** (28.5 g, 120 mmol, 63 % in 3 steps) as a white solid. mp 84–85 °C [lit 83–84 °C] [6]; ^{1}H NMR (400 MHz, $CDCl_3$) δ 12.7 (1H, s), 10.2 (1H, s), 3.92 (3H, s), 3.82 (3H, s), 2.49 (3H, s), 2.14 (3H, s); ^{13}C NMR (100 MHz, $CDCl_3$) δ 194.1, 168.0, 164.5, 162.2, 137.8, 122.0, 117.8, 114.6, 61.7, 52.4, 14.6, 8.2; IR (Neat) 3,447, 2,949, 1,734, 1,634, 1,581, 1,458, 1,305, 1,274, 1,210, 903, 791 cm^{-1}; HREIMS calcd for $C_{12}H_{14}O_5$ 238.0841, found 238.0842.

2-Benzyloxy-5-methoxycarbonyl-4-methoxy-3,6-dimethylbenzoic acid (13). To a solution of phenol **15** (22.5 g, 94.4 mmol) in DMF (150 mL) were added K_2CO_3 (45.6 g, 330 mmol, 3.5 equiv) and BnBr (14.6 mL, 123 mmol, 1.3 equiv) at room temperature under an argon atmosphere. After being stirred at the same temperature for 13 h, the reaction mixture was filtered through a pad of Celite®. The filtrate was diluted with EtOAc, and acidified with 3 M aq HCl. The organic layer was separated, and the aqueous layer was extracted with EtOAc twice. The combined organic layers were washed with brine twice, saturated aq $NaHCO_3$ and brine, dried with $MgSO_4$, and filtered. The filtrate was concentrated in vacuo to give benzyl ether as a colorless oil. The crude benzyl ether was used for next reaction without further purification.

To a suspension of crude aldehyde in *t*-BuOH (75 mL) and water (75 mL) were added 2-methyl-2-butene (50 mL), $NaClO_2$ (25.6 g, 283 mmol, 3.0 equiv) and $NaH_2PO_4 \cdot 2H_2O$ (44.2 g, 283 mmol, 3.0 equiv) at 0 °C. After being stirred at room temperature for 1.5 h, the organic layer was separated, and the aqueous layer was extracted with EtOAc twice. The combined organic layers were washed with brine, dried with $MgSO_4$, and filtered. The filtrate was concentrated in vacuo, and the resulting residue was purified by column chromatography on silica gel (eluted with hexane/EtOAc = 2:1). Recrystallization from CH_2Cl_2/hexane afforded carboxylic acid **13** (23.7 g, 68.8 mmol, 73 % in 2 steps) as a white solid. mp 96–97 °C [lit. 103–105 °C] [6]; ^{1}H NMR (400 MHz, $CDCl_3$) δ 7.31–7.44 (5H, m), 4.94 (2H, s), 3.95 (3H, s), 3.80 (3H, s), 2.33 (3H, s), 2.23 (3H, s); ^{13}C NMR (100 MHz, $CDCl_3$) δ 171.8, 168.2, 157.9, 156.5, 136.3, 132.6, 128.5, 128.3, 128.2, 126.0, 124.3, 123.5, 76.7, 61.8, 52.5, 16.9, 9.7; IR (Neat) 3,320, 2,950, 1,731, 1,580, 1,455, 1,312, 1,250, 1,104, 699 cm^{-1}; HREIMS calcd for $C_{19}H_{20}O_6$ 344.1260, found 344.1253.

2,5-Dimethylresorcinol (20).
To a solution of methyl 2,4-dihydroxy-3,6-dimethylbenzoate (10.0 g, 51.0 mmol) in ethylene glycol (50 mL) was added KOH (85 %, 234 mmol, 4.6 equiv) in water (50 mL) at 0 °C. After being stirred at 120 °C for 1 h, the reaction mixture was cooled to room temperature, diluted with EtOAc, and quenched with 6 M aq HCl at 0 °C. The organic layer was separated, and the aqueous layer was extracted with EtOAc twice. The combined organic layers were washed with saturated aq $NaHCO_3$ and brine, dried with $MgSO_4$, and filtered. The filtrate was concentrated in vacuo, and the resulting residue was recrystallized from EtOAc/hexane to afford diol **20** (6.40 g, 46.3 mmol, 91 %) as a white solid. mp 165–166 °C [lit. 161 °C] [38]; ^{1}H NMR (400 MHz, $CDCl_3$) δ 6.23 (2H, s), 4.59 (2H, s), 2.21 (3H, s), 2.10 (3H, s); ^{13}C NMR (100 MHz, CD_3OD) δ 157.1, 136.9, 109.0, 108.3, 27.3, 8.2; IR (Neat) 3,245, 1,630, 1,584, 1,521, 1,454, 1,416, 1,175, 1,075 cm^{-1}; HREIMS calcd for $C_8H_{10}O_2$ 138.0681, found 138.0711.

1,3-Dibenzyloxy-2,5-dimethylbenzene (12).
To a solution of diol **20** (5.00 g, 36.2 mmol) in DMF (100 mL) were added K_2CO_3 (22.5 g, 163 mmol, 4.5 equiv) and BnBr (9.47 mL, 79.6 mmol, 2.2 equiv) at room temperature under an argon atmosphere. After being stirred at the same temperature for 21 h, the reaction mixture was filtered through a pad of Celite®. The filtrate was diluted with EtOAc, and acidified with 3 M aq HCl. The organic layer was separated, and the aqueous layer was extracted with EtOAc twice. The combined organic layers were washed with brine twice, saturated aq $NaHCO_3$ and brine, dried with $MgSO_4$, and filtered. The filtrate was concentrated in vacuo, and the resulting residue was recrystallized from CH_2Cl_2/hexane to afford benzyl ether **12** (8.88 g, 28.6 mmol, 77 %) as a white needle. mp 92–93 °C [lit. 91 °C] [39]; ^{1}H NMR (400 MHz, $CDCl_3$) δ 7.28–7.46 (10H, m), 6.45 (2H, s), 5.06 (4H, s), 2.31 (3H, s), 2.19 (3H, s); ^{13}C NMR (100 MHz, $CDCl_3$) δ 157.5, 137.7, 136.2, 128.5, 127.7, 127.1, 112.5, 106.2, 70.3, 21.9, 8.4; IR (Neat) 2,920, 1,589, 1,453, 1,126, 695 cm^{-1}; HREIMS calcd for $C_{22}H_{22}O_2$ 318.1620, found 318.1634.

Methyl 4-benzyloxy-3-(2,4-dibenzyloxy-3,6-dimethylbenzoyl)-6-methoxy-2,5-di methylbenzoate (21).
To a solution of carboxylic acid **13** (3.90 g, 11.3 mmol) in dry CH_2Cl_2 (50 mL) were added benzyl ether **12** (7.20 g, 22.6 mmol, 2.0 equiv) and $(CF_3CO)_2O$ (7.86 mL, 56.5 mmol, 5.0 equiv) at 0 °C under an argon atmosphere. After being stirred at room temperature for 4.5 h, the reaction mixture was diluted with EtOAc, and quenched with 2 M aq NaOH. The organic layer was separated, and the aqueous layer was extracted with EtOAc twice. The combined organic layers were washed with brine, dried with $MgSO_4$, and filtered. The filtrate was concentrated in vacuo, and the resulting residue was purified by column chromatography on silica gel (eluted with hexane/EtOAc = 6:1) to afford benzophenone **21** (5.17 g, 8.03 mmol, 71 %) as a yellowish solid. Benzyl ether **12** (4.04 g, 12.7 mmol, 56 %) was recovered as a yellowish solid. mp 140–141 °C [lit. 109–112 °C] [6]; ^{1}H NMR (400 MHz, $CDCl_3$) δ 7.24–7.45 (11H, m), 7.14–7.19 (4H, m), 6.51 (1H, s), 5.06 (2H, s), 4.68 (2H, s), 4.62 (2H, s), 3.87 (3H, s), 3.71 (3H, s), 2.31 (3H, s), 2.11

(3H, s), 2.09 (3H, s), 2.00 (3H, s); ^{13}C NMR (100 MHz, $CDCl_3$) δ 197.2, 168.5, 159.4, 157.5, 157.1, 157.0, 137.7, 137.3, 137.2, 136.8, 134.2, 132.6, 128.6, 128.5, 128.09, 128.05 127.9, 127.5, 127.3, 127.1, 127.0, 126.6, 126.2, 123.0, 117.9, 110.6, 75.7, 70.0, 61.7, 52.1, 21.2, 16.8, 9.6, 9.4; IR (Neat) 2,947, 1,733, 1,654, 1,594, 1,569, 1,454, 1,314, 1,176, 1,125, 735, 697 cm^{-1}; HREIMS calcd for $C_{41}H_{40}O_7$ 644.2774, found 644.2767.

Methyl 3-(2,4-dihydroxy-3,6-dimethylbenoyl)-4-hydroxy-6-methoxy-2,5-dimethylbenzoate (11).
To a solution of benzyl ester **21** (7.00 g, 10.9 mmol) in EtOAc (120 mL) was added 5 % palladium on carbon (3.50 g, 50 wt%) at room temperature, and the flask was purged with hydrogen 7 times. After being stirred at the same temperature for 18 h, the reaction mixture was filtered through a pad of Celite®. The filtrate was concentrated in vacuo, and the resulting residue was recrystallized from CH_2Cl_2/hexane to afford phenol **11** (3.78 g, 10.1 mmol, 93 %) as a yellowish solid. mp 157–158 °C [lit. 164–165 °C] [6]; ^{1}H NMR (400 MHz, $CDCl_3$) δ 11.8 (1H, s), 7.79 (1H, s), 6.17 (1H, s), 5.30 (1H, s), 3.88 (3H, s), 3.80 (3H, s), 2.19 (3H, s), 2.14 (3H, s), 1.90 (3H, s), 1.80 (3H, s); ^{13}C NMR (100 MHz, CD_3OD) δ 202.5, 170.3, 166.4, 163.4, 158.5, 154.8, 141,5, 131.0, 129.6, 124.1, 118.3, 116.0, 112.3, 110.5, 62.5, 52.6, 22.5, 16.3, 9.2, 7.7; IR (Neat) 3,419, 2,969, 1,699, 1,602, 1,569, 1,267, 746 cm^{-1}; HREIMS calcd for $C_{20}H_{22}O_7$ 374.1366, found 374.1366.

Methyl 3-hydroxy-8-methoxy-1,4,6,9-tetramethyl-11-oxo-11*H*-dibenzo[*b*,*e*]-[1, 4]dioxepin-7-carboxylate (22).
To a suspension of benzophenone **11** (3.80 g, 10.2 mmol) in water (200 mL) were added K_2CO_3 (21.1 g, 153 mmol, 15 equiv) and $K_3[Fe(CN)_6]$ (7.69 g, 23.3 mmol, 2.3 equiv) at room temperature. After being stirred at the same temperature for 4 h, the reaction mixture was diluted with EtOAc, and quenched with 6 M aq HCl at 0 °C. The organic layer was separated, and the aqueous layer was extracted twice with EtOAc. The combined organic layers were washed with saturated aq $NaHCO_3$ and brine, dried with $MgSO_4$, and filtered. The filtrate was concentrated in vacuo, and the resulting residue was recrystallized from CH_2Cl_2/hexane to afford depsidone **22** (3.20 g, 8.59 mmol, 84 %) as an orange solid. mp 227–228 °C [lit. 220–221 °C] [6]; ^{1}H NMR (400 MHz, $CDCl_3$) δ 6.51 (1H, s), 5.42 (1H, s), 3.92 (3H, s), 3.74 (3H, s), 2.43 (3H, s), 2.37 (3H, s), 2.34 (3H, s), 2.28 (3H, s); ^{13}C NMR (100 MHz, CD_3OD) δ 169.1, 164.8, 162.5, 161.9, 153.9, 147.6, 146.0, 143.4, 127.7, 126.5, 123.0, 115.7, 114.5, 112.9, 62.6, 52.9, 21.2, 14.4, 10.0, 9.8; IR (Neat) 3,330, 2,930, 1,729, 1,699, 1,611, 1,590, 1,459, 1,413, 1,343, 1,272, 1,135 cm^{-1}; HREIMS calcd for $C_{20}H_{20}O_7$ 372.1209, found 372.1208.

Methyl 3,8-di(methoxymethoxy)-1,4,6,9-tetramethyl-11-oxo-11*H*-dibenzo-[*b*,*e*] [1, 4]dioxepine-7-carboxylate (10c).
To a suspension of methyl ether **22** (3.50 g, 9.40 mmol) in dry CH_2Cl_2 (30 mL) was added BCl_3 (1.0 M in CH_2Cl_2, 28.2 mL, 28.2 mmol, 3.0 equiv) at 0 °C under an argon atmosphere. After being stirred at room temperature for 4.5 h, the reaction mixture was diluted with EtOAc, and quenched with water at 0 °C. The organic

layer was separated, and the aqueous layer was extracted with EtOAc twice. The combined organic layers were washed with saturated aq $NaHCO_3$ and brine, dried with $MgSO_4$, and filtered. The filtrate was concentrated in vacuo to give phenol as a yellowish solid. The crude phenol was used for next reaction without further purification.

To a solution of crude diol in DMF (40 mL) were added DIPEA (16.4 mL, 94.0 mmol, 10 equiv) and MOMCl (3.57 mL, 47.0 mmol, 5.0 equiv) at room temperature under an argon atmosphere. After being stirred at the same temperature for 1 h, the reaction mixture was diluted with EtOAc, and quenched with 3 M aq HCl. The organic layer was separated and the aqueous layer was extracted with EtOAc twice. The combined organic layers were washed with brine twice, saturated aq $NaHCO_3$ and brine, dried with $MgSO_4$, and filtered. The filtrate was concentrated in vacuo, and the resulting residuc was purificd by column chromatography on silica gel (eluted with hexane/EtOAc = 3:1) to afford MOM ether **10c** (3.58 g, 8.01 mmol, 85 % in 2 steps) as a white solid.

mp 118–119 °C; ^{1}H NMR (400 MHz, $CDCl_3$) δ 6.80 (1H, s), 5.23 (2H, s), 4.93 (2H, s), 3.90 (3H, s), 3.53 (3H, s), 3.47 (3H, s), 2.47 (3H, s), 2.39 (3H, s), 2.33 (3H, s), 2.30 (3H, s); ^{13}C NMR (100 MHz, $CDCl_3$) δ 167.5, 162.9, 160.4, 158.9, 150.1, 146.5, 144.7, 142.1, 126.5, 125.6, 122.5, 115.8, 114.7, 113.1, 100.5, 94.2, 57.5, 56.3, 52.3, 21.3, 14.5, 10.5, 10.0; IR (Neat) 2,953, 1,735, 1,730, 1,607, 1,566, 1,458, 1,342, 1,281, 1,127, 977, 940, 759 cm^{-1}; HREIMS calcd for $C_{23}H_{26}O_9$ 446.1577, found 446.1559.

General procedure: Synthesis of benzyl ethers 10a and 10e.

To a suspension of the MOM ether **10c** in MeOH (10.0 mL/mmol) was added I_2 (1 wt/vol.% I_2/MeOH) at room temperature. After being stirred at 45 °C for 16 h, the reaction mixture was cooled to room temperature, diluted with EtOAc, and quenched with saturated aq $Na_2S_2O_3$. The organic layer was separated, and the aqueous layer was extracted with EtOAc twice. The combined organic layers were washed with saturated aq $NaHCO_3$ and brine, dried with $MgSO_4$, and filtered. The filtrate was concentrated in vacuo to give phenol as a white solid. The crude phenol was used for next reaction without further purification.

To a solution of the crude phenol in DMF were added K_2CO_3 and benzyl halide at room temperature under an argon atmosphere. After being stirred at the same temperature, the reaction mixture was filtered through a pad of Celite®. The filtrate was diluted with EtOAc, and acidified with 3 M aq HCl. The organic layer was separated, and the aqueous layer was extracted with EtOAc twice. The combined organic layers were washed with brine twice, saturated aq $NaHCO_3$ and brine, dried with $MgSO_4$, and filtered. The filtrate was concentrated in vacuo, and the resulting residue was purified to afford benzyl ethers **10a** and **10e**.

Methyl 8-benzyloxy-3-methoxymethoxy-1,4,6,9-tetramethyl-11-oxo-11*H*-dibenzo[*b,e*][1, 4]dioxepine-7-carboxylate (10a).
Conditions: (1) **10c** (150 mg, 336 μmol), I_2 (33.6 mg), MeOH (3.4 mL), 16 h. (2) BnBr (61.1 μL, 504 μmol, 1.5 equiv), K_2CO_3 (149 mg, 1.08 mmol, 3.2 equiv), DMF (2.0 mL), 10 h. Purification: flash column chromatography on silica gel (eluted with hexane/EtOAc = 4/1). Yield: benzyl ether **10a** (134 mg, 272 μmol, 81 % in 2 steps) as a white solid. mp 151–152 °C; ^{1}H NMR (400 MHz, $CDCl_3$) δ 7.31–7.41 (5H, m), 6.81 (1H, s), 5.23 (2H, s), 4.85 (2H, s), 3.82 (3H, s), 3.48 (3H, s), 2.48 (3H, s), 2.40 (3H, s), 2.35 (3H, s), 2.31 (3H, s); ^{13}C NMR (100 MHz, $CDCl_3$) δ 167.7, 163.0, 160.5, 158.9, 151.3, 146.4, 144.7, 142.1, 136.7, 128.5, 128.2, 127.9, 126.5, 125.6, 122.3, 115.8, 114.7, 113.1, 94.2, 76.6, 56.3, 52.4, 21.4, 14.4, 10.2, 10.0; IR (Neat) 2,952, 1,733, 1,606, 1,566, 1,455, 1,341, 1,281, 1,204, 1,130, 1,063, 979 cm^{-1}; HREIMS calcd for $C_{28}H_{28}O_8$ 492.1784, found 492.1793.

Methyl 8-(4-chlorobenzyloxy)-3-methoxymethoxy-1,4,6,9-tetramethyl-11-oxo-11*H*-dibenzo[*b,e*][1, 4]dioxepine-7-carboxylate (10e).
Conditions: (1) **10c** (3.20 g, 717 mmol), I_2 (717 mg), MeOH (72 mL), 16 h. (2) (4-Cl)BnCl (2.30 g, 14.3 mmol, 2.0 equiv), K_2CO_3 (4.95 g, 35.9 mmol, 5.0 equiv), DMF (70 mL), 11 h. Purification: recrystallization from CH_2Cl_2/hexane after column chromatography on silica gel (eluted with hexane/EtOAc = 9/1). Yield: benzyl ether **10e** (2.40 g, 4.55 mmol, 63 % in 2 steps) as a white solid. mp 129–130 °C; ^{1}H NMR (400 MHz, $CDCl_3$) δ 7.35 (2H, d, J = 8.8 Hz), 7.32 (2H, d, J = 8.8 Hz), 6.81 (1H, s), 5.23 (2H, s), 4.82 (2H, s), 3.80 (3H, s), 3.47 (3H, s), 2.48 (3H, s), 2.39 (3H, s), 2.34 (3H, s), 2.29 (3H, s); ^{13}C NMR (100 MHz, $CDCl_3$) δ 167.6, 162.8, 160.4, 158.9, 151.1, 146.4, 144.7, 142.1, 135.2, 134.0, 129.1, 128.6, 126.4, 125.6, 122.2, 115.8, 114.6, 113.1, 94.1, 75.6, 56.3, 52.4, 21.3, 14.4, 10.2, 10.0; IR (Neat) 2,952, 1,735, 1,606, 1,454, 1,342, 1,282, 1,204, 1,130, 1,063, 980, 759 cm^{-1}; HREIMS calcd for $C_{28}H_{27}ClO_8$ 526.1394, found 526.1380.

Methyl 8-methoxy-3-methoxymethoxy-1,4,6,9-tetramethyl-11-oxo-11*H*-dibenzo[*b,e*][1, 4]dioxepine-7-carboxylate (10d).
To a solution of phenol **22** (60 mg, 161 μmol) in DMF (1.0 mL) were added DIPEA (84.1 μL, 483 μmol, 3.0 equiv) and MOMCl (20.2 μL, 242 μmol, 1.5 equiv) at room temperature under an argon atmosphere. After being stirred at the same temperature for 1.5 h, the reaction mixture was diluted with EtOAc, and quenched with 1 M aq HCl. The organic layer was separated, and the aqueous layer was extracted with EtOAc twice. The combined organic layers were washed with brine twice, saturated aq $NaHCO_3$ and brine, dried with $MgSO_4$, and filtered. The filtrate was concentrated in vacuo, and the resulting residue was purified by column chromatography on silica gel (eluted with hexane/EtOAc = 3:1) to afford MOM ether **10d** (52.3 mg, 126 μmol, 78 %) as a white solid. mp 142–143 °C; ^{1}H NMR (400 MHz, $CDCl_3$) δ 6.80 (1H, s), 5.23 (2H, s), 3.92 (3H, s), 3.74 (3H, s), 3.47 (3H, s), 2.48 (3H, s), 2.38 (3H, s), 2.34 (3H, s), 2.28 (3H, s); ^{13}C NMR (100 MHz, $CDCl_3$) δ 167.7, 162.9, 160.5, 158.9, 152.5, 146.2, 144.7, 142.1, 126.1, 125.4, 122.0, 115.8, 114.7, 113.1, 94.2, 62.1, 56.3, 52.4, 21.3, 14.3, 10.0, 9.9; IR (Neat)

2,931, 1,735, 1,606, 1,566, 1,459, 1,414, 1,344, 1,282, 1,129, 1,063 cm^{-1}; HREIMS calcd for $C_{22}H_{24}O_8$ 416.1471, found 416.1480.

General procedure: Synthesis of benzyl alcohols 28.
To a solution of depsidones **10** in dry THF was added $NaBH_4$ at 0 °C under an argon atmosphere. After being stirred at the optimal temperature, the reaction mixture was diluted with EtOAc, and quenched with 3 M aq HCl at 0 °C. The organic layer was separated, and the aqueous layer was extracted with EtOAc twice. The combined organic layers were washed with saturated aq $NaHCO_3$ and brine, dried with $MgSO_4$, and filtered. The filtrate was concentrated in vacuo, and the resulting residue was purified to afford benzyl alcohols **28**.

Methyl 2-benzyloxy-4-hydroxy-5-(2-hydroxymethyl-5-methoxymethoxy-3,6-dimethyl-phenoxy)-3,6-dimethylbenzoate (28a).
Conditions: **10a** (40.0 mg, 81.2 μmol), $NaBH_4$ (77.4 mg, 2.03 mmol, 25 equiv), dry THF (1.0 mL), 0 °C, 74 h. Purification: column chromatography on silica gel (eluted with hexane/EtOAc = 3/1). Yield: benzyl alcohol **28a** (20.6 mg, 41.4 μmol, 52 %) as a white solid, depsidone **10a** (10.2 mg, 20.7 μmol, 26 % recovered). mp 147–148 °C; ^{1}H NMR (400 MHz, $CDCl_3$) δ 7.30–7.43 (5H, m), 6.74 (1H, s), 5.19 (1H, d, J = 6.8 Hz), 5.13 (1H, d, J = 6.8 Hz, g), 5.04 (1H, d, J = 10.6 Hz), 4.90 (1H, d, J = 11.2 Hz), 4.85 (1H, d, J = 11.2 Hz), 4.72 (1H, d, J = 10.6 Hz), 3.81 (3H, s), 3.47 (3H, s), 2.40 (3H, s), 2.18 (3H, s), 2.13 (3H, s), 1.80 (3H, s); ^{13}C NMR (100 MHz, $CDCl_3$) δ 168.6, 156.2, 155.3, 150.6, 147.6, 140.7, 137.3, 136.1, 128.4, 128.0, 127.9, 124.4, 122.0, 121.5, 118.9, 115.7, 111.8, 94.7, 76.4, 57.3, 56.1, 52.1, 19.3, 13.8, 9.4, 8.9; IR (Neat) 3,423, 2,953, 2,926, 1,728, 1,611, 1,585, 1,456, 1,287, 1,112, 1,056 cm^{-1}; HRESITOFMS calcd for $C_{28}H_{32}O_8Na$ $[M+Na]^+$ 519.1995, found 519.1993.

Methyl 4-hydroxy-3-(2-hydroxymethyl-5-methoxymethoxy-3,6-dimethyl-phenoxy)-6-methoxy-methoxy-2,5-dimethylbenzoate (28c).
Conditions: **10c** (30.0 mg, 66.6 μmol), $NaBH_4$ (37.8 mg, 999 μmol, 15 equiv), dry THF (1.0 mL), 0 °C, 94 h. Purification: flash column chromatography on silica gel (eluted with hexane/EtOAc = 4/1). Yield: benzyl alcohol **28c** (6.2 mg, 13.7 μmol, 21 %) as a white solid, and depsidone **10c** (9.9 mg, 22.2 μmol, 33 % recovered). mp 156–157 °C; ^{1}H NMR (400 MHz, $CDCl_3$) δ 6.71 (1H, s), 5.17 (1H, d, J = 6.8 Hz), 5.11 (1H, d, J = 6.8 Hz), 5.00 (1H, d, J = 10.8 Hz), 4.94 (2H, s), 4.68 (1H, d, J = 10.8 Hz), 3.88 (3H, s), 3.53 (3H, s), 3.45 (3H, s), 2.38 (3H, s), 2.14 (3H, s), 2.12 (3H, s), 1.75 (3H, s); ^{13}C NMR (100 MHz, $CDCl_3$) δ 168.5, 156.1, 155.2, 149.3, 147.6, 140.8, 136.0, 124.4, 121.9, 121.4, 119.0, 115.7, 111.7, 100.4, 94.6, 57.5, 57.2, 56.1, 52.1, 19.3, 13.9, 9.7, 8.9; IR (Neat) 3,424, 2,953, 2,929, 1,728, 1,612, 1,584, 1,462, 1,287, 1,209, 1,156, 1,113, 1,089, 1,057, 971, 755 cm^{-1}; HRESITOFMS calcd for $C_{23}H_{30}O_9Na$ $[M+Na]^+$ 473.1788, found 473.1786.

Methyl 4-hydroxy-3-(2-hydroxymethyl-5-methoxymethoxy-3,6-dimethyl-phenoxy)-6-methoxy-2,5-dimethylbenzoate (28d).
Conditions: **10d** (8.0 mg, 19.2 μmol), $NaBH_4$ (20.0 mg, 529 μmol, 28 equiv), dry THF (1.0 mL), rt, 44 h. Purification: flash column chromatography on silica gel

(eluted with hexane/EtOAc = 2/1). Yield: benzyl alcohol **28d** (6.6 mg, 15.7 μmol, 82 %) as a white solid. mp 145–146 °C; ^{1}H NMR (400 MHz, $CDCl_3$) δ 6.73 (1H, s), 5.18 (1H, d, $J = 6.4$ Hz), 5.13 (1H, d, $J = 6.4$ Hz), 5.01 (1H, d, $J = 10.8$ Hz), 4.70 (1H, $J = 10.8$ Hz, d), 3.90 (3H, s), 3.74 (3H, s), 3.46 (3H, s), 2.40 (3H, s), 2.14 (3H, s), 2.11 (3H, s), 1.79 (3H, s); ^{13}C NMR (100 MHz, $CDCl_3$) δ 168.6, 156.1, 155.2, 152.0, 147.5, 140.5, 136.1, 124.2, 121.9, 121.1, 118.5, 115.8, 111.7, 94.6, 62.1, 57.4, 56.1, 52.1, 19.4, 13.7, 9.1, 8.9; IR (Neat) 3,417, 2,925, 1,729, 1,611, 1,464, 1,416, 1,288, 1,207, 1,112, 1,089, 1,058 cm^{-1}; HRESITOFMS calcd for $C_{22}H_{28}O_8Na$ $[M+Na]^+$ 443.1682, found 443.1689.

Methyl 2-(4-chlorobenzyloxy)-4-hydroxy-5-(2-hydroxymethyl-5-methoxy-methoxy-3,6-dimethyl-phenoxy)-3,6-dimethylbenzoate (28e).
Conditions: **10e** (680 mg, 1.29 mmol), $NaBH_4$ (1.95 g, 51.6 mmol, 40 equiv), dry THF (103 mL), 30 °C, 103 h. Purification: Recrystallization from CH_2Cl_2/hexane after flash column chromatography on silica gel (eluted with hexane/EtOAc = 4/1). Yield: benzyl alcohol **28e** (474 mg, 89.2 μmol, 69 %) as a white solid, and depsidone **10e** (64.4 mg, 122 μmol, 10 % recovered). mp 159–160 °C; ^{1}H NMR (400 MHz, $CDCl_3$) δ 7.34 (4H, s), 6.71 (1H, s), 5.17 (1H, d, $J = 6.8$ Hz), 5.12 (1H, d, $J = 6.8$ Hz), 5.00 (1H, d, $J = 10.8$ Hz), 4.85 (1H, d, $J = 10.8$ Hz), 4.81 (1H, d, $J = 10.8$ Hz), 4.67 (1H, d, $J = 10.8$ Hz), 3.78 (3H, s), 3.45 (3H, s), 2.37 (3H, s), 2.13 (3H, s), 2.09 (3H, s), 1.76 (3H, s); ^{13}C NMR (100 MHz, $CDCl_3$) δ 168.6, 156.1, 155.2, 150.3, 147.7, 140.8, 136.1, 135.7, 133.7, 129.2, 128.5, 124.4, 121.9, 121.3, 118.7, 115.6, 111.7, 94.6, 75.5, 57.2, 56.1, 52.1, 19.3, 13.8, 9.4, 8.9; IR (Neat) 3,422, 2,952, 2,927, 1,726, 1,611, 1,586, 1,492, 1,460, 1,289, 1,209, 1,113, 1,089, 1,057, 980, 755 cm^{-1}; HRESITOFMS calcd for $C_{28}H_{31}ClO_8Na$ $[M+Na]^+$ 553.1605, found 553.1606.

Methyl 4-benzyloxy-2-(4-chlorobenzyloxy)-5-(2-hydroxymethyl-5-methoxy-methoxy-3,6-dimethyl-phenoxy)-3,6-dimethylbenzoate (30).
To a solution of phenol **28e** (530 mg, 99.8 μmol) in acetone (4.0 mL) were added K_2CO_3 (552 mg, 3.99 mmol, 4.0 equiv) and BnBr (154 μL, 1.30 mmol, 1.3 equiv) at room temperature under an argon atmosphere. After being stirred at the same temperature for 5 h, the reaction mixture was diluted with EtOAc, and quenched with 1 M aq HCl. The organic layer was separated, and the aqueous layer was extracted with ethyl acetate twice. The combined organic layers were washed with saturated aq $NaHCO_3$ and brine, dried with $MgSO_4$, and filtered. The filtrate was concentrated in vacuo, and the resulting residue was purified by column chromatography on silica gel (eluted with hexane/EtOAc = 9:1) to afford benzyl ether **30** (572 mg, 92.1 μmol, 92 %) as a white solid. mp 108–109 °C; ^{1}H NMR (400 MHz, $CDCl_3$) δ 7.35 (4H, s), 7.24–7.27 (3H, m), 7.01–7.04 (2H, m), 6.69 (1H, s), 5.17 (1H, d, $J = 7.0$ Hz), 5.14 (1H, d, $J = 7.0$ Hz), 4.86 (2H, s), 4.68 (1H, d, $J = 11.6$ Hz, l), 4.55 (1H, d, $J = 7.0$ Hz), 4.52 (1H, d, $J = 7.0$ Hz), 4.12 (1H, d, $J = 11.6$ Hz), 3.84 (3H, s), 3.45 (3H, s), 2.26 (3H, s), 2.24 (3H, s), 2.14 (3H, s), 1.98 (3H, s); ^{13}C NMR (100 MHz, $CDCl_3$) δ 168.2, 155.7, 155.4, 149.4, 148.4, 146.1, 137.0, 136.2, 135.6, 133.9, 129.1, 128.6, 128.2, 127.9, 127.3, 126.4, 124.9, 124.8, 124.2, 115.4, 111.8, 94.6, 75.5, 74.8, 57.1, 56.0, 52.3, 19.1, 13.9, 10.4, 9.5; IR (Neat) 3,522, 2,952, 1,731, 1,610, 1,580, 1,492,

1,444, 1,367, 1,337, 1,284, 1,207, 1,115, 1,091, 1,059, 983, 753 cm^{-1}; HREIMS calcd for $C_{35}H_{37}ClO_8$ 620.2177, found 620.2150.

Methyl 4-benzyloxy-2-(4-chlorobenzyloxy)-5-(3-hydroxy-2,5,6-trimethyl-phenoxy)-3,6-dimethyl-benzoate (31).

To a solution of benzyl alcohol **30** (460 mg, 741 μmol) in dry CH_2Cl_2 (7.6 mL) were added TFA (0.4 mL) and Et_3SiH (2.32 mL, 14.8 mmol, 20 equiv) at—30 °C under an argon atmosphere, and the mixture was stirred at 0 °C for 30 min. To the reaction mixture were added methanol (8.0 mL) and 6 M aq HCl (1.6 mL) at 0 °C. After being stirred at 50 °C for 19 h, the reaction mixture was cooled to room temperature, diluted with EtOAc, and quenched with saturated aq $NaHCO_3$. The organic layer was separated, and the aqueous layer was extracted with EtOAc twice. The combined organic layers were washed with brine, dried with $MgSO_4$, and filtered. The filtrate was concentrated in vacuo, and the resulting residue was purified by column chromatography on silica gel (eluted with hexane/EtOAc = 5:1) to afford phenol **31** (367 mg, 654 μmol, 88 %) as a colorless oil. ^{1}H NMR (400 MHz, $CDCl_3$) δ 7.34 (4H, s), 7.25–7.28 (3H, m), 7.06–7.09 (2H, m), 6.36 (1H, s), 4.85 (2H, s), 4.67 (1H, brs), 4.58 (1H, d, J = 11.2 Hz), 4.49 (1H, d, J = 11.2 Hz), 3.82 (3H, s), 2.14 (3H, s), 2.10 (3H, s), 2.08 (3H, s), 1.98 (3H, s), 1.93 (3H, s); ^{13}C NMR (100 MHz, $CDCl_3$) δ 168.6, 155.0, 151.9, 149.1, 148.9, 146.2, 137.0, 135.7, 135.5, 133.8, 129.1, 128.6, 128.1, 127.7, 127.3, 125.9, 124.6, 124.4, 119.4, 112.3, 111.8, 75.4, 74.5, 52.3, 19.9, 13.9, 12.5, 10.2, 9.4; IR (Neat) 3,443, 2,950, 2,925, 1,729, 1,598, 1,493, 1,454, 1,407, 1,367, 1,285, 1,212, 1,099, 755 cm^{-1}; HREIMS calcd for $C_{33}H_{33}ClO_6$ 560.1966, found 560.1942.

Methyl 4-benzyloxy-2-(4-chlorobenzyloxy)-5-(3-methoxy-2,5,6-trimethyl-phenoxy)-3,6-dimethyl-benzoate (32).

To a solution of phenol **31** (500 mg, 891 μmol) in DMF (4.0 mL) were added K_2CO_3 (690 mg, 4.99 mmol, 5.6 equiv) and MeI (128 μL, 2.05 mmol, 2.3 equiv) at room temperature under an argon atmosphere. After being stirred at the same temperature for 8 h, the reaction mixture was diluted with EtOAc, and quenched with 1 M aq HCl. The organic layer was separated, and the aqueous layer was extracted with EtOAc twice. The combined organic layers were washed with brine twice, saturated aq $NaHCO_3$ and brine, dried with $MgSO_4$, and filtered. The filtrate was concentrated in vacuo, and the resulting residue was purified by column chromatography on silica gel (eluted with hexane/EtOAc = 9:1) to afford methyl ether **32** (469 mg, 815 μmol, 91 %) as a colorless oil. ^{1}H NMR (400 MHz, $CDCl_3$) δ 7.34 (4H, s), 7.24–7.26 (3H, m), 7.05–7.07 (2H, m), 6.41 (1H, s), 4.85 (2H, s), 4.57 (1H, d, J = 11.2 Hz), 4.47 (1H, d, J = 11.2 Hz), 3.82 (3H, s), 3.74 (3H, s), 2.15 (3H, s), 2.13 (3H, s), 2.09 (3H, s), 2.00 (3H, s), 1.93 (3H, s); ^{13}C NMR (100 MHz, $CDCl_3$) δ 168.5, 155.9, 154.9, 149.0, 148.8, 146.2, 137.1, 135.7, 135.0, 133.8, 129.1, 128.6, 128.1, 127.6, 127.3, 125.9, 124.5, 124.4, 119.6, 114.4, 107.7, 75.4, 74.4, 55.7, 52.2, 20.3, 13.9, 12.6, 10.2, 9.4; IR (Neat) 2,950, 2,927, 1,731, 1,613, 1,599, 1,583, 1,492, 1,338, 1,283, 1,206, 1,125, 754 cm^{-1}; HREIMS calcd for $C_{34}H_{35}ClO_6$ 574.2122, found 574.2133.

4-Benzyloxy-2-(4-chlorobenzyloxy)-5-(3-methoxy-2,5,6-trimethyl-phenoxy)-3,6-dimethylbenzoic acid (6b).
To a solution of methyl ester **32** (145 mg, 252 μmol) in dioxane (1.0 mL) were added 2 M aq NaOH (2.0 mL, 4.00 mmol, 16 equiv) and ethylene glycol (0.5 mL) at room temperature. After being stirred at 200 °C under microwave irradiation for 40 min, the reaction mixture was cooled to room temperature, diluted with EtOAc, and quenched with 3 M aq HCl. The organic layer was separated, and the aqueous layer was extracted with EtOAc twice. The combined organic layers were washed with brine, dried with $MgSO_4$, and filtered. The filtrate was concentrated in vacuo, and the resulting residue was purified by column chromatography on silica gel (eluted with hexane/EtOAc = 4:1) to afford carboxylic acid **6b** (118 mg, 211 μmol, 84 %) as an orange oil. ^{1}H NMR (400 MHz, $CDCl_3$) δ 7.23–7.32 (7H, m), 7.03–7.05 (2H, m), 6.41 (1H, s), 4.85 (2H, s), 4.54 (1H, d, J = 11.2 Hz), 4.44 (1H, d, J = 11.2 Hz), 3.74 (3H, s), 2.26 (3H, s), 2.14 (3H, s), 2.06 (3H, s), 1.99 (3H, s), 1.92 (3H, s); ^{13}C NMR (100 MHz, $CDCl_3$) δ 173.1, 155.8, 154.9, 149.1, 148.9, 146.3, 137.0, 135.2, 135.0, 133.9, 129.5, 128.6, 128.0, 127.6, 127.2, 125.5, 124.9, 124.5, 119.5, 114.4, 107.7, 75.6, 74.3, 55.7, 20.3, 14.0, 12.6, 10.2, 9.4; IR (Neat) 3,380, 2,926, 1,725, 1,699, 1,614, 1,599, 1,580, 1,492, 1,455, 1,365, 1,219, 1,126, 1,092 cm^{-1}; HREIMS calcd for $C_{33}H_{33}ClO_6$ 560.1966, found 560.1941.

Methyl 4-benzyloxy-2-(4-chlorobenzyloxy)-5-(4-formyl-3-methoxy-2,5,6-trimethylphenoxy)-3,6-dimethylbenzoate (34).
To a solution of methyl ether **32** (30.0 mg, 52.2 μmol) in dry CH_2Cl_2 (1.0 mL) were added Cl_2CHOMe (46.1 μL, 522 μmol, 10 equiv) and AgOTf (29.5 mg, 115 μmol, 2.2 equiv) at—40 °C under an argon atmosphere. After being stirred at 0 °C for 30 min, the reaction mixture was quenched with saturated aq $NaHCO_3$ at 0 °C. After being stirred at room temperature for 30 min, the reaction mixture was filtered through a pad of Celite®. The organic layer was separated, and the aqueous layer was extracted with EtOAc twice. The combined organic layers were washed with brine, dried with $MgSO_4$, and filtered. The filtrate was concentrated in vacuo to give desired aldehyde **34** and deprotected phenol. The crude mixture was used for the next reaction without further purification.

To a solution of the crude mixture in DMF (1.0 mL) were added K_2CO_3 (36.1 mg, 261 μmol, 5.0 equiv) and (4-Cl)BnCl (16.8 mg, 104 μmol, 2.0 equiv) at room temperature under an argon atmosphere. After being stirred at the same temperature for 17 h, the reaction mixture was diluted with EtOAc, and quenched with 1 M aq HCl. The organic layer was separated, and the aqueous layer was extracted with EtOAc twice. The combined organic layers were washed with brine twice, saturated aq $NaHCO_3$ and brine, dried with $MgSO_4$, and filtered. The filtrate was concentrated in vacuo, and the resulting residue was purified by flash column chromatography on silica gel (eluted with hexane/EtOAc = 9:1) to afford aldehyde **34** (15.7 mg, 26.2 μmol, 50 % in 2 steps) as a colorless oil. ^{1}H NMR (400 MHz, $CDCl_3$) δ 10.4 (1H, s), 7.35 (4H, s), 7.21–7.22 (3H, m), 6.92–6.95 (2H, m), 4.87 (2H, s), 4.55 (1H, d, J = 11.4 Hz), 4.51 (1H, d, J = 11.4 Hz), 3.85 (3H, s), 3.63 (3H, s), 2.37 (3H, s), 2.21 (3H, s), 2.12 (3H, s), 2.09 (3H, s), 1.93 (3H, s); ^{13}C NMR

(100 MHz, $CDCl_3$) δ 192.4, 168.2, 162.2, 159.1, 149.6, 148.3, 145.2, 138.8, 136.4, 135.5, 133.9, 129.1, 128.6, 128.2, 127.8, 126.5, 126.2, 124.7, 124.6, 124.5, 123.7, 119.3, 15.5, 14.2, 62.9, 52.4, 15.9, 13.8, 12.8, 10.2, 9.7; IR (Neat) 2,928, 1,731, 1,685, 1,565, 1,454, 1,337, 1,284, 1,206, 1,108 cm^{-1}; HRFABMS calcd for $C_{35}H_{36}ClO_7$ $[M + H]^+$ 603.2150, found 603.2128.

Methyl 4-benzyloxy-2-(4-chlorobenzyloxy)-5-(4-carboxy-3-methoxy-2,5,6-trim ethylphenoxy)-3,6-dimethylbenzoate (35).
To a solution of aldehyde **xx** (15.0 mg, 24.9 μmol) in *t*-BuOH (0.5 mL) and water (0.5 mL) were added 2-methyl-2-butene (0.5 mL), $NaClO_2$ (22.5 mg, 249 μmol, 10 equiv) and $NaH_2PO_4 \cdot 2H_2O$ (38.8 mg, 249 μmol, 10 equiv) at 0 °C. After being stirred at room temperature for 14 h, the organic layer was separated, and the aqueous layer was extracted with EtOAc twice. The combined organic layers were washed with brine, dried with $MgSO_4$, and filtered. The filtrate was concentrated in vacuo, and the resulting residue was purified by column chromatography on silica gel (eluted with hexane/EtOAc = 2:3) to afford carboxylic acid **35** (9.8 mg, 15.8 μmol, 64 %) as a colorless oil. ^1H NMR (400 MHz, $CDCl_3$) δ 7.35 (4H, s), 7.25–7.30 (3H, m), 6.97–6.98 (2H, m), 4.87 (2H, s), 4.56 (1H, d, $J = 11.2$ Hz), 4.17 (1H, d, $J = 11.2$ Hz), 3.85 (3H, s), 3.67 (3H, s), 2.20 (3H, s), 2.17 (3H, s), 2.13 (3H, s), 2.07 (3H, s), 1.96 (3H, s); ^{13}C NMR (100 MHz, $CDCl_3$) δ 170.8, 168.3, 156.4, 154.4, 149.5, 148.5, 145.5, 136.6, 135.6, 134.3, 133.9, 129.1, 128.6, 128.3, 127.8, 126.7, 126.2, 124.6, 124.5, 124.3, 122.5, 119.5, 75.5, 74.3, 62.1, 52.3, 16.9, 13.9, 13.1, 10.2, 10.1; IR (Neat) 3,429, 2,950, 2,931, 1,732, 1,703, 1,600, 1,455, 1,285, 1,208, 1,105 cm^{-1}; HREIMS calcd for $C_{35}H_{35}ClO_8$ 618.2020, found 618.2030.

Methyl 4-benzyloxy-2-methoxy-3,5,6-trimethylbenzoate (41).
To a solution of aldehyde **15** (2.00 g, 10.1 mmol) in dry THF (20 mL) were added Et_3N (1.69 mL, 12.1 mmol, 1.2 equiv) and ClCOOEt (1.16 mL, 12.1 mmol, 1.2 equiv) at 0 °C under an argon atmosphere. After being stirred at the same temperature for 40 min, the reaction mixture was filtered through a pad of Celite®. To filtrate was added $NaBH_4$ (1.53 g, 40.4 mmol, 4.0 equiv) in water (20 mL) at 0 °C. After being stirred at room temperature for 2.5 h, the reaction mixture was diluted with EtOAc, and acidified with 3 M aq HCl. The organic layer was separated, and the aqueous layer was extracted with EtOAc twice. The combined organic layers were washed with saturated aq $NaHCO_3$ and brine, dried with $MgSO_4$, and filtered. The filtrate was concentrated in vacuo to give trimethylbenzene as a white solid. The crude trimethylbenzene was used for next reaction without further purification.

To a solution of crude phenol in DMF (20 mL) were added K_2CO_3 (4.19 g, 30.3 mmol, 3.0 equiv) and BnBr (1.68 mL, 14.1 mmol, 1.4 equiv) at room temperature under an argon atmosphere. After being stirred at the same temperature for 11 h, the reaction mixture was filtered through a pad of Celite®. The filtrate was diluted with EtOAc, and acidified with 3 M aq HCl. The organic layer was separated, and the aqueous layer was extracted with EtOAc twice. The combined organic layers were washed with brine twice, saturated aq $NaHCO_3$ and brine, dried with $MgSO_4$, and filtered. The filtrate was concentrated in vacuo, and the resulting residue was purified by column chromatography on silica gel (eluted with hexane/EtOAc = 5:1) to the afford benzyl ether **41** (3.01 g, 9.56 mmol, 95 % in 2 steps) as a yellowish oil. ^{1}H NMR (400 MHz, $CDCl_3$) δ 7.35–7.48 (5H, m), 4.75 (2H, s), 3.93 (3H, s), 3.75 (3H, s), 2.21 (3H, s), 2.18 (3H, s), 2.17 (3H, s); ^{13}C NMR (100 MHz, $CDCl_3$) δ 169.1, 156.9, 153.7, 137.0, 132.3, 128.2, 127.8, 127.5, 125.9, 125.2, 121.9, 74.2, 61.4, 51.8, 16.4, 12.3, 9.4; IR (Neat) 2,948, 1,731, 1,577, 1,455, 1,327, 1,284, 1,200, 1,177, 1,105, 1,005 cm^{-1}; HREIMS calcd for $C_{19}H_{22}O_4$ 314.1518, found 314.1499.

4-Benzyloxy-2-methoxy-3,5,6-trimethylbenzoic acid (42).
To a solution of methyl ester **41** (300 mg, 954 μmol) in dioxane (1.2 mL) were added 2 M aq NaOH (2.4 mL, 4.80 mmol, 5.0 equiv) and ethylene glycol (0.8 mL) at room temperature. After being stirred at 180 °C under microwave irradiation for 50 min, the reaction mixture was cooled to room temperature, diluted with EtOAc, and quenched with 3 M aq HCl. The organic layer was separated, and the aqueous layer was extracted with EtOAc twice. The combined organic layers were washed with brine, dried with $MgSO_4$, and filtered. The filtrate was concentrated in vacuo, and the resulting residue was purified by column chromatography on silica gel (eluted with hexane/EtOAc = 1:1) to afford carboxylic acid **42** (251 mg, 836 μmol, 88 %) as a white solid. mp 124–125 °C; ^{1}H NMR (400 MHz, $CDCl_3$) δ 7.36–7.47 (5H, m), 4.77 (2H, s), 3.81 (3H, s), 2.34 (3H, s), 2.23 (3H, s), 2.22 (3H, s); ^{13}C NMR (100 MHz, $CDCl_3$) δ 173.0, 157.7, 154.2, 137.1, 133.6, 128.6, 128.2, 127.8, 126.7, 123.9, 122.4, 74.5, 62.1, 16.9, 12.7, 9.8; IR (Neat) 2,934, 1,699, 1,578, 1,455, 1,322, 1,221,1,106, 696 cm^{-1}; HREIMS calcd for $C_{18}H_{20}O_4$ 300.1362, found 300.1347.

Benzyl 4-hydroxy-2-methoxy-3,5,6-trimethylbenzoate (8).
To a solution of benzyl ether **42** (3.50 g, 11.7 mmol) in EtOAc (30 mL) was added 5 % palladium on carbon (2.0 g, 57 wt%) at room temperature, and the flask was purged with hydrogen 7 times. After being stirred at room temperature for 16 h, the reaction mixture was filtered through a pad of Celite®. The filtrate was concentrated in vacuo to give phenol as a white solid. The crude phenol was used for next reaction without further purification.

To a solution of crude carboxylic acid in DMF (30 mL) were added $KHCO_3$ (3.21 g, 32.1 mmol, 2.7 equiv) and BnBr (1.53 mL, 12.8 mmol, 1.1 equiv) at room temperature under an argon atmosphere. After being stirred at the same temperature for 9.5 h, the reaction mixture was filtered through a pad of Celite®. The filtrate was diluted with EtOAc, and acidified with 3 M aq HCl. The organic layer was separated, and the aqueous layer was extracted with EtOAc twice. The combined organic layers were washed with brine twice, saturated aq $NaHCO_3$ and brine, dried with $MgSO_4$, and filtered. The filtrate was concentrated in vacuo, and the resulting residue was purified by column chromatography on silica gel (eluted with hexane/EtOAc = 5:1) to afford benzyl ester **8** (3.29 g, 10.9 mmol, 94 % in 2 steps) as an orange solid. mp 61–62 °C; ^{1}H NMR (400 MHz, $CDCl_3$) δ 7.31–7.45 (5H, m), 5.36

(2H, s), 4.78 (1H, s), 3.64 (3H, s), 2.14 (6H, s), 2.11 (3H, s); ^{13}C NMR (100 MHz, $CDCl_3$) δ 169.0, 153.8, 153.7, 135.8, 132.3, 128.5, 128.4, 128.2, 121.5, 118.3, 114.9, 66.9, 61.9, 16.6, 11.6, 8.8; IR (Neat) 3,466, 2,943, 1,721, 1,584, 1,456, 1,291, 1,190, 1,108 cm^{-1}; HREIMS calcd for $C_{18}H_{20}O_4$ 300.1362, found 300.1369.

4-Acetoxy-2-methoxy-3,5,6-trimethylbenzoic acid (14).
To a solution of phenol **8** (1.50 g, 4.99 mmol) in dry CH_2Cl_2 (10 mL) were added Ac_2O (0.567 mL, 5.99 mmol, 1.2 equiv), Et_3N (2.09 mL, 15.0 mmol, 3.0 equiv) and DMAP (61.0 mg, 499 μmol, 0.10 equiv) at room temperature under an argon atmosphere. After being stirred at the same temperature for 15 h, the reaction mixture was diluted with EtOAc, and quenched with 3 M aq HCl. The organic layer was separated, and the aqueous layer was extracted with EtOAc twice. The combined organic layers were washed with saturated aq $NaHCO_3$ and brine, dried with $MgSO_4$, and filtered. The filtrate was concentrated in vacuo, to give acetate as a white solid. The crude acetate was used for next reaction without further purification.

To a solution of crude benzyl ester in EtOAc (30 mL) was added 5 % palladium on carbon (800 mg, 53 wt%) at room temperature, and the flask was purged with hydrogen 7 times. After being stirred at the same temperature for 24 h, the reaction mixture was filtered through a pad of Celite®. The filtrate was concentrated in vacuo, and the resulting residue was recrystallized from CH_2Cl_2/hexane to afford carboxylic acid **14** (1.09 g, 4.33 mmol, 87 % in 2 steps) as a white solid. mp 158–159 °C; ^{1}H NMR (400 MHz, $CDCl_3$) δ 3.81 (3H, s), 2.37 (3H, s), 2.31 (3H, s), 2.09 (3H, s), 2.06 (3H, s); ^{13}C NMR (100 MHz, $CDCl_3$) δ 173.3, 168.6, 153.6, 149.7, 132.9, 126.3, 125.5, 121.7, 62.3, 20.4, 16.8, 12.7, 9.7; IR (Neat) 3,265, 2,942, 1,754, 1,737, 1,579, 1,462, 1,370, 1,200, 1,097 cm^{-1}; HREIMS calcd for $C_{13}H_{16}O_5$ 252.0998, found 252.0997.

4-Benzyloxycarbonyl-3-methyl-2,5,6-trimethylphenyl 4-hydroxy-2-methoxy-3,5,6-trimethylbenzoate (7).
To a solution of carboxylic acid **14** (1.09 g, 4.32 mmol, 1.2 equiv) in dry toluene (30 mL) were added phenol **8** (1.08 g, 3.60 mmol) and $(CF_3CO)_2O$ (2.00 mL, 14.4 mmol, 4.0 equiv) at room temperature under an argon atmosphere. After being stirred at the same temperature for 12 h, the reaction mixture was quenched with 2 M aq NaOH. The organic layer was separated, and the aqueous layer was extracted twice with EtOAc. The combined organic layers were washed with brine, dried with $MgSO_4$, and filtered. The filtrate was concentrated in vacuo to give ester as a colorless oil. The crude ester was used for next reaction without further purification.

To a solution of crude acetate in MeOH (10 mL) was added K_2CO_3 (746 mg, 5.40 mmol, 1.5 equiv) at room temperature. After being stirred at the same temperature for 2 h, the reaction mixture was diluted with EtOAc, and quenched with 1 M aq HCl. The organic layer was separated, and the aqueous layer was extracted twice with EtOAc. The combined organic layers were washed with saturated aq $NaHCO_3$ and brine, dried with $MgSO_4$, and filtered. The filtrate was concentrated in vacuo, and the resulting residue was recrystallized from CH_2Cl_2/hexane to afford phenol **7** (1.59 g, 3.23 mmol, 90 % in 2 steps) as a white solid. mp 172–173 °C; ^{1}H NMR (400 MHz, $CDCl_3$) δ 7.33–7.47 (5H, m), 5.39 (2H, s), 4.90 (1H, s), 3.79 (3H, s), 3.70 (3H, s), 2.36 (3H, s), 2.23 (3H, s), 2.22 (3H, s), 2.193 (3H, s), 2.188 (3H, s), 2.17 (3H, s); ^{13}C NMR (100 MHz, $CDCl_3$) δ 168.4, 166.8, 154.6, 154.5, 153.5, 149.7, 135.6, 133.3, 132.6, 128.50, 128.48, 128.3, 127.1, 125.7, 122.1, 120.4, 118.7, 114.5, 67.0, 62.1, 61.9, 17.1, 16.7, 12.9, 11.8, 10.1, 9.0; IR (Neat) 3,481, 2,942, 1,736, 1,733, 1,576, 1,463, 1,456, 1,286, 1,159, 1,097, 1,078 cm^{-1}; HRFABMS calcd for $C_{29}H_{32}O_7Na$ $[M+Na]^+$ 515.2046, found 515.2076.

Preparation of ester 43.
To a solution of carboxylic acid **6b** (70.0 mg, 125 μmol, 1.2 equiv) in dry toluene (2.0 mL) were added phenol **7** (51.2 mg, 104 μmol) and $(CF_3CO)_2O$ (579 μL, 4.16 mmol, 40 equiv) at room temperature under an argon atmosphere. After being stirred at 80 °C for 9 h, the reaction mixture was cooled at room temperature, and quenched with 2 M aq NaOH. The organic layer was separated, and the aqueous layer was extracted with EtOAc twice. The combined organic layers were washed with brine, dried with $MgSO_4$, and filtered. The filtrate was concentrated in vacuo, and the resulting residue was purified by flash column chromatography on silica gel (eluted with hexane/EtOAc = 9:1) to afford ester **43** (105 mg, 101 μmol, 97 %) as a white solid. mp 85–86 °C; ^{1}H NMR (400 MHz, $CDCl_3$) δ 7.46–7.48 (2H, m), 7.26–7.40 (10H, m), 7.06–7.08 (2H, m), 6.44 (1H, s), 5.40 (2H, s), 4.94 (2H, s), 4.60 (1H, d, J = 11.0 Hz), 4.49 (1H, d, J = 11.0 Hz), 3.78 (3H, s), 3.76 (3H, s), 3.71 (3H, s), 2.41 (3H, s), 2.35 (3H, s), 2.25 (3H, s), 2.21 (3H, s), 2.18 (3H, s), 2.171 (3H, s), 2.168 (3H, s), 2.14 (3H, s), 2.10 (3H, s), 2.05 (3H, s), 1.98 (3H, s); ^{13}C NMR (100 MHz, $CDCl_3$) δ 168.3, 166.2, 165.8, 155.9, 154.8, 154.3, 153.6, 150.0, 149.8, 149.5, 149.4, 146.4, 136.9, 135.7, 135.4, 135.1, 133.8, 133.4, 132.7, 129.2, 128.54, 128.50, 128.3, 128.1, 127.7, 127.4, 127.2, 126.5, 126.1, 125.7, 125.2, 124.9, 124.7, 122.5, 122.2, 119.5, 114.4, 107.7, 75.7, 74.4, 67.0, 62.1, 62.0, 55.7, 20.4, 17.3, 16.7, 14.5, 13.5, 13.0, 12.6, 10.6, 10.4, 10.2, 9.5; IR (Neat) 2,939, 1,738, 1,578, 1,463, 1,322, 1,278, 1,150, 1,122, 1,095, 1,076 cm^{-1}; HRFABMS calcd for $C_{62}H_{63}ClO_{12}Na$ $[M+Na]^+$ 1,057.3906, found 1,057.3905.

Preparation of aldehyde 44.
To a solution of ester **43** (260 mg, 251 μmol) in dry CH_2Cl_2 (3.0 mL) were added Cl_2CHOMe (225 μL, 2.51 mmol, 10 equiv) and AgOTf (142 mg, 552 μmol, 2.2 equiv) at—40 °C under an argon atmosphere. After being stirred at 0 °C for 30 min, the reaction mixture was quenched with saturated aq $NaHCO_3$ at 0 °C. After being stirred at room temperature for 30 min, the reaction mixture was filtered through a pad of Celite®. The organic layer of the filtrate was separated, and the aqueous layer was extracted with EtOAc twice. The combined organic layers were washed with brine, dried with $MgSO_4$, and filtered. The filtrate was concentrated in vacuo to give the mixture of desired aldehyde **44** and deprotected phenol. The crude mixture was used for next reaction without further purification.

To a solution of crude mixture in DMF (3.0 mL) were added K_2CO_3 (104 mg, 753 μmol, 3.0 equiv) and (4-Cl)BnCl (60.6 mg, 377 μmol, 1.5 equiv) at room temperature under an argon atmosphere. After being stirred at the same temperature for 14 h, the reaction mixture was diluted with EtOAc, and quenched with 1 M aq HCl. The organic layer was separated, and the aqueous layer was extracted with EtOAc twice. The combined organic layers were washed with brine twice, saturated aq $NaHCO_3$ and brine, dried with $MgSO_4$, and filtered. The filtrate was concentrated in vacuo, and the resulting residue was purified by flash column chromatography on silica gel (eluted with hexane/EtOAc = 9:1) to afford aldehyde **44** (144 mg, 136 μmol, 54 % in 2 steps) as a colorless oil. ^{1}H NMR (400 MHz, $CDCl_3$) δ 10.4 (1H, s), 7.46–7.47 (2H, m), 7.31–7.40 (8H, m), 7.23–7.24 (2H, m), 6.96–6.98 (2H, m), 5.40 (2H, s), 4.97 (2H, s), 4.60 (1H, d, J = 11.4 Hz), 4.55 (1H, d, J = 11.4 Hz), 3.79 (3H, s), 3.71 (3H, s), 3.65 (3H, s), 2.47 (3H, s), 2.39 (3H, s), 2.36 (3H, s), 2.25 (3H, s), 2.22 (3H, s), 2.20 (3H, s), 2.18 (3H, s), 2.15 (3H, s), 2.14 (3H, s), 2.11 (3H, s), 1.99 (3H, s); ^{13}C NMR (100 MHz, $CDCl_3$) δ 192.4, 168.3, 166.1, 165.5, 162.2, 159.0, 154.3, 153.7, 150.4, 150.0, 149.4, 149.0, 145.5, 138.9, 136.3, 135.7, 135.3, 134.0, 133.5, 132.7, 129.2, 128.59, 128.56, 128.53, 128.3, 128.2, 127.9, 127.4, 126.6, 126.5, 126.1, 125.7, 125.29, 125.25, 124.79, 124.75, 123.8, 122.4, 122.2, 119.3, 75.8, 74.2, 67.1, 63.0, 62.1, 62.0, 17.3, 16.7, 15.9, 14.5, 13.5, 13.0, 12.9, 10.7, 10.4, 10.2, 9.8; IR (Neat) 2,932, 1,740, 1,684, 1,573, 1,460, 1,323, 1,278, 1,150, 1,111, 755 cm^{-1}; HRESITOFMS calcd for $C_{63}H_{63}ClO_{13}Na$ $[M + Na]^+$ 1085.3855, found 1085.3860.

Preparation of carboxylic acid 9b.

To a solution of aldehyde **44** (26.0 mg, 24.4 μmol) in *t*-BuOH (1.0 mL) and water (1.0 mL) were added 2-methyl-2-butene (0.5 mL), $NaClO_2$ (22.1 mg, 244 μmol, 10 equiv) and $NaH_2PO_4 \cdot 2H_2O$ (38.1 mg, 244 μmol, 10 equiv) at 0 °C. After being stirred at room temperature for 4 h, the organic layer was separated, and the aqueous layer was extracted with EtOAc twice. The combined organic layers were washed with brine, dried with $MgSO_4$, and filtered. The filtrate was concentrated in vacuo, and the resulting residue was purified by flash column chromatography on silica gel (eluted with hexane/EtOAc = 1:1) to afford carboxylic acid **9b** (21.9 mg, 20.3 μmol, 83 %) as a yellowish oil. ^{1}H NMR (400 MHz, $CDCl_3$) δ 7.46–7.48 (2H, m), 7.28–7.40 (10H, m), 6.99–7.01 (2H, m), 5.40 (2H, s), 4.97 (2H, s), 4.62 (1H, d, J = 11.4 Hz), 4.51 (1H, d, J = 11.4 Hz, t), 3.79 (3H, s), 3.71 (3H, s), 3.70 (3H, s), 2.46 (3H, s), 2.37 (3H, s), 2.26 (3H, s), 2.22 (6H, s), 2.19 (3H, s), 2.18 (3H, s), 2.16 (3H, s), 2.12 (6H, s), 2.01 (3H, s); ^{13}C NMR (100 MHz, $CDCl_3$) δ 171.2, 168.3, 166.2, 165.6, 156.3, 154.4, 153.7, 150.2, 150.0, 149.5, 149.2, 145.7, 136.4, 135.7, 135.3, 134.2, 134.0, 133.5, 132.7, 129.2, 128.6, 128.5, 128.34, 128.32, 127.9, 127.4, 126.7, 126.6, 126.1, 125.7, 125.3, 125.2, 124.8, 124.3, 122.9, 122.4, 122.2, 119.4, 75.8, 74.3, 67.1, 62.1, 62.0, 17.3, 16.9, 16.7, 14.5, 13.5, 13.2, 13.0, 10.7, 10.4, 10.24, 10.17; IR (Neat) 2,939, 1,738, 1,733, 1,600, 1,575, 1,462, 1,323, 1,279, 1,152, 1,097, 755 cm^{-1}; HRESITOFMS calcd for $C_{63}H_{63}ClO_{14}Na$ $[M + Na]^+$ 1,101.3804, found 1,101.3816.

Preparation of ester 45.
To a solution of carboxylic acid **9b** (110 mg, 102 μmol) in dry toluene (5.0 mL) were added phenol **8** (36.7 mg, 122 μmol, 1.2 equiv) and $(CF_3CO)_2O$ (568 μL, 4.08 mmol, 40 equiv) at room temperature under an argon atmosphere. After being stirred at 80 °C for 11 h, the reaction mixture was cooled at room temperature, and quenched with 2 M aq NaOH. The organic layer was separated, and the aqueous layer was extracted with EtOAc twice. The combined organic layers were washed with brine, dried with $MgSO_4$, and filtered. The filtrate was concentrated in vacuo, and the resulting residue was purified by flash column chromatography on silica gel (eluted with hexane/EtOAc = 6:1) to afford ester **45** (105 mg, 77.3 μmol, 76 %) as a white solid. mp 106–107 °C; ^{1}H NMR (500 MHz, $CDCl_3$) δ 7.46–7.48 (4H, m), 7.21–7.40 (13H, m), 7.01–7.02 (2H, m), 5.402 (2H, s), 5.398 (2H, s), 4.97 (2H, s), 4.63 (1H, d, J = 11.2 Hz), 4.52 (1H, d, J = 11.2 Hz), 3.79 (3H, s), 3.72 (3H, s), 3.71 (3H, s), 3.68 (3H, s), 2.48 (3H, s), 2.37 (3H, s), 2.26 (3H, s), 2.25 (3H, s), 2.23 (3H, s), 2.22 (6H, s), 2.21 (3H, s), 2.19 (3H, s), 2.18 (3H, s), 2.16 (3H, s), 2.15 (3H, s), 2.12 (3H, s), 2.05 (3H, s); ^{13}C NMR (125 MHz, $CDCl_3$) δ 168.33, 168.30, 166.4, 166.2, 165.6, 156.2, 154.7, 154.4, 153.7, 150.2, 150.0, 149.6, 149.5, 149.2, 145.7, 136.4, 135.7, 135.3, 134.0, 133.53, 133.49, 132.71, 132.69, 129.2, 128.59, 128.58, 128.55, 128.4, 128.3, 127.9, 127.43, 127.36, 126.8, 126.6, 126.1, 125.71, 125.69, 125.31, 125.26, 124.8, 124.1, 123.8, 122.4, 122.20, 122.19, 119.7, 75.8, 74.4, 67.1, 62.1, 62.0, 61.9, 17.3, 17.2, 16.8, 16.7, 14.5, 13.5, 13.2, 13.01, 12.99, 10.7, 10.44, 10.38, 10.3; IR (Neat) 3,011, 2,940, 1,735, 1,600, 1,576, 1,461, 1,323, 1,280, 1,149, 1,096, 1,077, 754 cm^{-1}; HRESITOFMS calcd for $C_{81}H_{81}ClO_{17}Na$ $[M + Na]^+$ 1383.5060, found 1383.5070.

Synthesis of thielocin B1 (1)
To a solution of benzyl ester **45** (5.0 mg, 3.7 μmol) in EtOH (0.5 mL) and EtOAc (0.5 mL) was added 10 % palladium on carbon (10 mg, 200 wt%) at room temperature, and the flask was purged with hydrogen 7 times. After being stirred at the same temperature for 15 min, the reaction mixture was filtered through a pad of Celite®. The filtrate was concentrated in vacuo, and the resulting residue was purified by preparative TLC (eluted with $CHCl_3$/MeOH = 3:1) to afford thielocin B1 (**1**) (1.9 mg, 2.0 μmol, 54 %) as a white solid. mp > 300 °C; ^{1}H NMR (400 MHz, CD_3OD) δ 3.832 (3H, s n), 3.825 (3H, s), 3.80 (3H, s), 3.73 (3H, s), 2.39 (3H, s), 2.38 (3H, s), 2.36 (3H, s), 2.284 (3H, s), 2.276 (3H, s), 2.22 (3H, s), 2.21 (3H, s), 2.20 (6H, s), 2.18 (3H, s), 2.14 (6H, s), 2.11 (3H, brs), 2.01 (3H, s); ^{13}C NMR (100 MHz, CD_3OD) δ 171.3, 168.4, 167.8, 161.0, 158.4, 156.2, 155.7, 153.23, 153.22, 150.9, 148.5, 148.4, 140.1, 136.7, 136.6, 134.50, 134.47, 131.63, 131.58, 128.5, 128.1, 127.1, 125.9, 125.1, 124.8, 123.3, 122.3, 120.2, 111.9, 62.8, 62.6, 62.01, 61.99, 17.6, 17.4, 17.1, 16.5, 13.4, 13.3, 10.6, 10.54, 10.51, 10.4, 8.6; IR (Neat) 3,430, 2,918, 1,736, 1,656, 1,572, 1,460, 1,160, 1,093, 1,074 cm^{-1}; HRESITOFMS calcd for $C_{53}H_{59}O_{17}$ $[M + H]^+$ 967.3752, found 967.3751.

Table 2.4 PPI inhibitory activity of synthesized **1** for PAC3 homodimer

Synthesized **1** (nM)	Fluorescence intensity ($\times 10^6$)				Intensity ratio (%)
	1st	2nd	3rd	Mean	
Background	0.273	0.258	0.258	0.263	0
Control	4.40	4.39	4.30	4.36	100
625	0.549	0.658	0.538	0.582	8
313	0.702	0.783	0.698	0.727	11
156	0.897	0.986	0.869	0.917	16
78.1	1.47	3.31	1.49	2.09	45
39.1	2.36	2.38	2.40	2.38	52
19.5	3.20	3.20	3.20	3.20	72
9.77	3.61	3.79	3.76	3.72	84
4.88	3.94	4.20	4.09	4.07	93
2.44	4.11	4.33	4.30	4.25	97
1.22	4.25	4.29	4.36	4.30	98

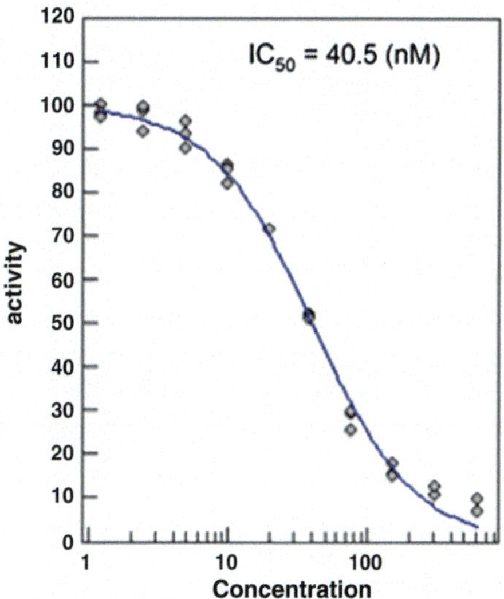

IC$_{50}$ value of synthesized thielocin B1 (1) for PAC3 homodimer and PAC1/PAC2 heterodimer.

The evaluation of PPIs inhibition of synthesized **1** for PAC3 (mKG_N-PAC3: 10 nM and mKG_C-PAC3: 25 nM, respectively) homodimer and PAC1/PAC2 (PAC1-mKG_N: 79 nM and mKG_C-PAC2: 120 nM, respectively) were performed as previously reported [2]. Aliquot of 3 μL of a protein solution were dispensed into a 384-well plate using a Multidrop Combi dispenser, and 0.1 μL of **1** dissolved in DMSO was added using a Multi-dispenser ADS-348-8. After incubation at room temperature for 1 h, 3 μL of the other protein solution was

Table 2.5 PPI inhibitory activity of synthesized **1** for PAC1/PAC2

Synthesized **1** (nM)	Fluorescence intensity ($\times 10^6$)				Intensity ratio (%)
	1st	2nd	3rd	Mean	
Background	0.650	0.567	0.563	0.594	0
Control	6.59	6.44	5.97	6.33	100
625	5.95	6.07	6.19	6.07	95
313	6.20	6.36	6.30	6.29	99
156	6.22	6.39	6.32	6.31	100
78.1	6.34	6.41	6.36	6.37	101
39.1	6.28	6.32	4.89	5.83	91
19.5	6.34	6.32	6.27	6.31	100
9.77	5.98	6.06	6.17	6.07	95
4.88	6.16	6.23	6.22	6.20	98
2.44	6.02	6.13	6.22	6.12	96
1.22	5.99	6.07	6.44	6.17	97

dispensed. After incubation for 17 h, fluorescence was measured using an EnVision reader. The results were summarized in Tables 2.4 and 2.5.

Protein expression and purification.
The cDNA encoding PAC3 was cloned into the pRSF-Duet1 vector (Novagen) to express PAC3 with an N-terminal hexahistidine tag, followed by an enterokinase cleavage site. Using this construct, the protein was expressed in *Escherichia coli* BL21 (DE3) Codon-Plus cells, which were grown in M9 minimal medium containing [^{15}N]NH_4Cl (1 g/L) using a previously described protocol [40]. The uniformly ^{15}N-labeled protein was purified using a Ni Sepharose 6 Fast Flow (GE Healthcare Bio-Sciences). Subsequently, the histidine tag was cleaved by enterokinase treatment, followed by anion exchange chromatography using a Resource Q column (GE Healthcare).

NMR experiment for chemical shift perturbation.
^{15}N-labeled PAC3 (0.2 mM) sample, dissolved in PBS (pH 6.8) containing 10 % D_2O (v/v), 1 mM DTT, 1 mM EDTA, and 0.01 % NaN_3, was titrated with a 10 mM methanol-d_4 solution of **1** (0.1, 0.2, 0.4, or 0.8 mM) or the solvent alone as a reference control. All NMR data were acquired at 303 K using a Bruker DMX500 spectrometer equipped with a cryogenic probe. For ^{1}H–^{15}N HSQC measurements, spectra were recorded at a ^{1}H observation frequency of 500 MHz with 128 (t_1) × 1024 (t_2) complex points and 8 scans per t_1 increment. ^{1}H chemical shifts were referenced to DSS (0 ppm), while ^{13}C chemical shifts were indirectly referenced to DSS using the absolute frequency ratios. Chemical shift changes were quantified as $[(\Delta\delta_H)^2 + (0.2\ \Delta\delta_N)^2]^{1/2}$, where $\Delta\delta_H$ and $\Delta\delta_N$ are the observed chemical shift changes for ^{1}H and ^{15}N, respectively. The data were processed using NMRPipe [41] software and analyzed with SPARKY 3 [42] and CCPNMR [43] software. Conventional 3D NMR experiments [44] were carried out for assignments of the HSQC peaks originating from PAC3 homodimer.

References

1. Matsumoto K, Tanaka K, Matsutani S, Sakazaki R, Hinoo H, Uotani N, Tanimoto T, Kawamura Y, Nakamoto S, Yoshida T. J Antibiot. 1995;48:106–12.
2. Hashimoto J, Watanabe T, Seki T, Karasawa S, Izumikawa M, Seki T, Iemura S, Natsume T, Nomura N, Goshima N, Miyawaki A, Takagi M, Shin-ya K. J Biomol Screen. 2009;14:970–9.
3. Yashiroda H, Mizushima T, Okamoto K, Kameyama T, Hayashi H, Kishimoto T, Niwa S, Kasahara M, Kurimoto E, Sakata E, Takagi K, Suzuki A, Hirano Y, Murata S, Kato K, Yamane T, Tanaka K. Nat Struct Mol Biol. 2008;15:228–36.
4. Hirano Y, Hayashi H, Iemura S, Hendil KV, Niwa S, Kishimoto T, Kasahara M, Natsume T, Tanaka K, Murata S. Mol Cell. 2006;24:977–84.
5. Sperotto E, van Klink GPM, de Vries JG, van Koten G. The synthesis of 2,2′,6,6′-tetrasubstituted diphenyl ether by the coupling reaction with transition metals has been reported by Koten; however the yield was quite low even in the simple substrates. Tetrahedron 2010; 66:9009–20.
6. Sala T, Sargent SV. J Chem Soc Perkin Trans I. 1981; 877–82.
7. Smith WE. J Org Chem. 1972;37:3972–3.
8. Lindgren BO, Nilsson T. Acta Chem Scand. 1972;27:888–90.
9. Kraus GA, Taschener MJ. J Org Chem. 1980;45:1174–5.
10. Bal BS, Childers WE, Pinnick HW. Tetrahedron. 1981;37:2091–6.
11. Sala T, Sargent SV. J Chem Soc Chem Commun. 1978; 1043–1044.
12. Comber MF, Sargent MV, Skelton BW, White AH. The intermediate has been isolated in some cases. J Chem Soc Perkin Trans I. 1989; 441–447.
13. We tried the isolation of intermediate **25** under low reaction temperature. Although compound **22** and unknown product were monitored on TLC and crude ^{1}H NMR, these two products were converged to **22** during the purification.
14. Keith JM. Tetrahedron Lett. 2004;45:2739–42.
15. Jaffé HH. Chem Rev. 1953;53:191–261.
16. Tanaka H, Miyoshi H, Chuang Y-C, Ando Y, Takahashi T. Angew Chem Int Ed. 2007;46:5934–7.
17. West CT, Donnelly SJ, Kooistra DA, Doyle MP. J Org Chem. 1973;38:2675–81.
18. Kappe CO. Angew Chem Int Ed. 2004;43:6250–84.
19. Meyer V. Chem Ber. 1894;27:510–2.
20. Meyer V, Sudborough J. Chem Ber. 1894;27:3146–53.
21. Vilsmeier A, Haack A. Ber Deutsch Chem Ges. 1927;60:119–22.
22. Reimer K. Ber Dtsch Chem Ges. 1876;9:423–4.
23. Reimer K, Tiemann F. Ber Dtsch Chem Ges. 1876;9:824–8.
24. Reimer K, Tiemann F. Ber Dtsch Chem Ges. 1876;9:1268–78.
25. Wynberg H. Chem Rev. 1960;60:169–84.
26. Rieche A, Gross H, Höft E. Chem Ber. 1960;93:88–94.
27. Gross H, Rieche A, Mattey G. Chem Ber. 1963;96:308–19.
28. The further investigation of newly developed formylation is mentioned in Chapter 4.
29. Fletcher S, Gunning PT. Removal of the benzyl group using chelating effects under acidic conditions has been reported. Tetrahedron Lett. 2008; 49:4817–19.
30. Hojo M, Masuda R, Okada E. Tetrahedron Lett. 1987;28:6199–200.
31. Tepczewski JJ, Callahan MP, Neghbors JD, Wiemer DF. J Am Chem Soc. 2009;131:14630–1.
32. Mente NR, Neghbors JD, Wiemer DF. J Org Chem. 2008;73:7963–70.
33. Minami N, Kijima S. Chem Pharm Bull. 1979;27:1490–4.
34. Parish RC, Stock LM. J Org Chem. 1965;30:927–9.
35. Corey EJ, Nicolaou KC. J Am Chem Soc. 1974;96:5614–6.
36. Gerlash H, Thalmann A. Helv Chim Acta. 1974;57:2661–3.
37. Molecular Operating Environment (MOE), ver. 2012, Chemical Computing Group Inc., 2012.

38. Cresp TM, Djura P, Sargent MV, Elix JA, Engkaninan U, Murphy DPH. Aust J Chem. 1975;28:2417–34.
39. Sargent MV, Vogel P, Elix JA, Ferguson BA. Aust J Chem. 1976;29:2263–9.
40. Yagi-Utsumi M, Matsuo K, Yanagisawa K, Gekko K, Kato K. Int J Alzheimers Dis. 2011; 925073. doi:10.4061/2011/925073.
41. Delaglio F, Grzesiek S, Vuister GW, Zhu G, Pfeifer J, Bax A. J Biomol NMR. 1995;6:277–93.
42. Goddard TD, Kneller DG. SPARKY 3. University of California, San Francisco.
43. Vranken WF, Boucher W, Stevens TJ, Fogh RH, Pajon A, Llinas M, Ulrich EL, Markley JL, Ionides J, Laue ED. Protein. 2005;59:687–96.
44. Uekusa Y, Mimura S, Sasakawa H, Kurimoto E, Sakata E, Serve O, Yagi H, Tokunaga F, Iwai K, Kato K. Biomol NMR Assign. 2012;6:177–80.

Chapter 3
Synthesis of Spin-Labeled Thielocin B1 and Observation of Paramagnetic Relaxation Enhancement Effects

3.1 Introduction

As mentioned in Chap. 2, the first total synthesis of thielocin B1 (**1**) was completed by our group. The results of NMR chemical shift perturbation experiments with proteasome-assembling chaperone 3 (PAC3) in the presence of synthesized **1** and in silico docking studies indicated that compound **1** affects the dissociation of PAC3 homodimer to monomeric PAC3 (Fig. 3.1). However, further studies were required to validate the inhibition mechanism because it was possible that the observed changes in the chemical shifts of these amino acid residues could have been induced by structural changes in PAC3 in the presence of **1**. With this in mind, a spin-labeling experiment was planned using NMR analysis to obtain further details about the binding site of **1**. In this chapter, the synthesis of spin-labeled **1** is described in detail and the inhibition mechanism of thielocin B1 further characterized based on the results of paramagnetic relaxation enhancement (PRE) experiments using the spin-labeled molecular probe and additional in silico docking studies.

3.2 Strategy

Based on the pose of thielocin B1 proposed by an in silico docking study, it was envisaged that the best area to extract information pertaining to the way in which thielocin B1 interacts with PAC3 homodimer would be around area $\underline{A}$, as shown in Fig. 3.2. In this way, the front of one protomer and the back of another could be located on the same side of PAC3 homodimer, so that when the binding site of **1** is set as the former, area $\underline{A}$ would correspond to the latter. In this case, one would expect to see a distinct decrease in the intensities of the NMR signals belonging to the residues in area $\underline{A}$ of only the **1**/PAC3 homodimer complex because this area

K. Ohsawa, *Total Synthesis of Thielocin B1 as a Protein–Protein Interaction Inhibitor of PAC3 Homodimer*, Springer Theses, DOI 10.1007/978-4-431-55447-9_3

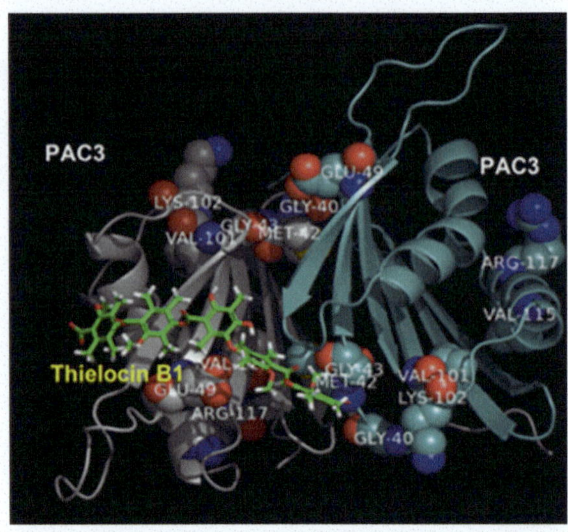

Fig. 3.1 Proposed pose for the interaction between **1** and PAC3 homodimer based on the NMR chemical shift perturbation experiment. Residues exhibiting significant changes in their chemical shifts are shown as the CPK model

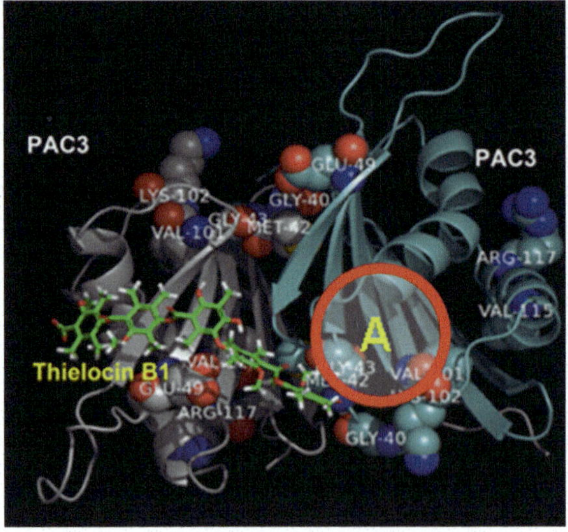

Fig. 3.2 The area planned for the extraction of information from the spin-labeling experiment. Area A is located on the back of the binding site of **1** with monomeric PAC3

would be too remote in a **1**/monomeric PAC3 complex to be influenced by the presence of thielocin B1.

Spin-labeled thielocin B1 (**2a**) was designed and synthesized based on the proposed pose of thielocin B1, as shown in Fig. 3.3. To obtain the desired information, it was decided that the wing to the left-hand side of **1**, which was closer to the interface of PAC3 homodimer, should be labeled. In particular, the hydroxyl group located on the wing to the left-hand side of the molecule was identified as the best position to attach a spin-labeling reagent because the carboxyl groups in **1** could behave as an anchor towards PAC3 homodimer. Subsequent consideration of

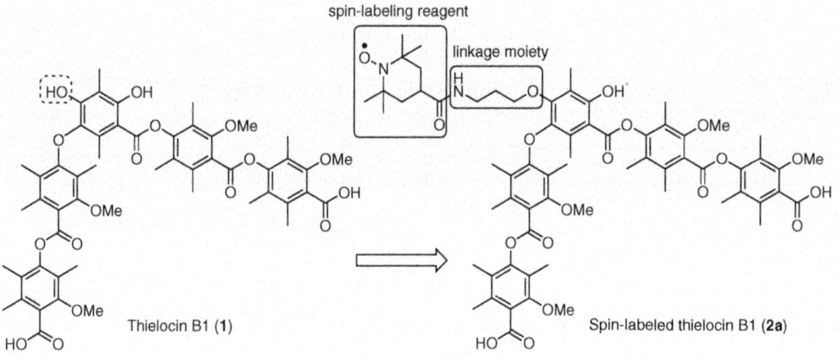

Fig. 3.3 The design of spin-labeled thielocin B1 (**2a**)

the linkage moiety led to the selection of a 3-aminopropyl ether linker. Using a longer linker was also considered, but long and flexible carbon chains can make it difficult to extract meaningful topological information. Additionally, when the amino group is extricated to be amidated with a spin-labeling reagent such as commercially available 4-carboxy-TEMPO, the elimination of the linker could be avoided by intramolecular cyclization.

The synthetic strategy for the construction of **2a** is shown in Scheme 3.1. The direct and selective etherification of **1** would be difficult in the presence of two hydroxyl and two carboxyl groups, and a suitably selective procedure would

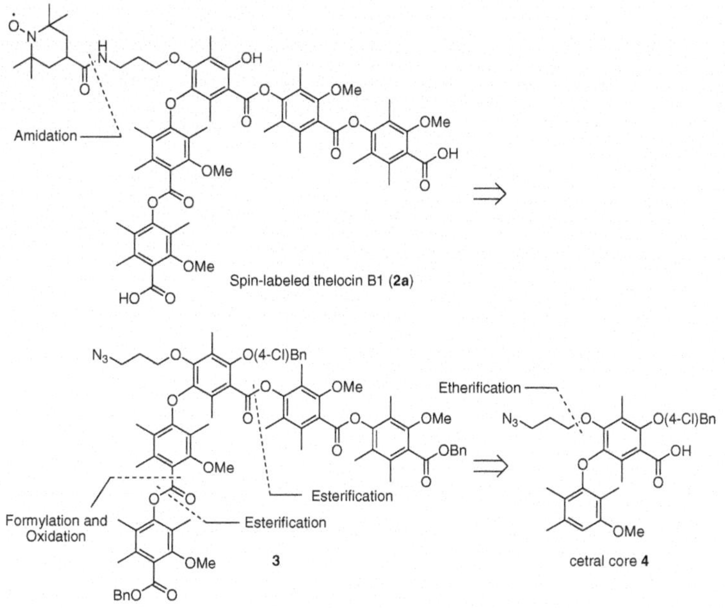

Scheme 3.1 Retrosynthesis of spin-labeled thielocin B1 (**2a**)

therefore be required to allow for the etherification of a synthetic intermediate of **1** and the introduction of the aminopropyl linker. Given that the nitroxyl radical in this spin-labeling reagent could complicate the analysis of the product by NMR, it was decided that compound **3** would be spin-labeled via an amidation reaction with 4-carboxy-TEMPO following the removal of all of the benzyl groups. Compound **3** could be prepared from the central core **4** according to the method used in the total synthesis of **1**. The linkage moiety in **4** could be introduced by the etherification of the phenolic hydroxyl group.

3.3 Synthesis of Spin-Labeling Precursor

3.3.1 Modified Synthesis of Central Core

The synthetic route used for the construction of the central core was initially optimized, as shown in Scheme 3.2. Methylation of phenol **5**, followed by the selective demethylation of methyl ether with BCl_3 and benzylation of the resulting phenol with benzyl chloride afforded **6**. The lactone moiety in **6** was selectively reduced with $NaBH_4$ using the optimized conditions described in Chap. 2, Table 2.1. Unfortunately, however, the desired bisalcohol **7** was obtained as an impure mixture containing several unknown and inseparable products in 27 % yield, and **6** was recovered in 62 % yield following a scalable synthesis. Following a review of several alternative reducing agents, it was established that $LiBH_4$ prepared in situ from $NaBH_4$ and LiBr was particularly effective for completing the reaction to provide **7** quantitatively. The success of $LiBH_4$ in this reduction reaction was attributed to its high solubility in THF and the high affinity of the Li^+ cation for the carbonyl oxygen of the lactone group, which would have enhanced its reactivity [1]. Subsequent reduction of the resulting benzylic alcohol with Et_3SiH and TFA [2] gave compound **8**, which was subjected to an etherification reaction using 1-azido-

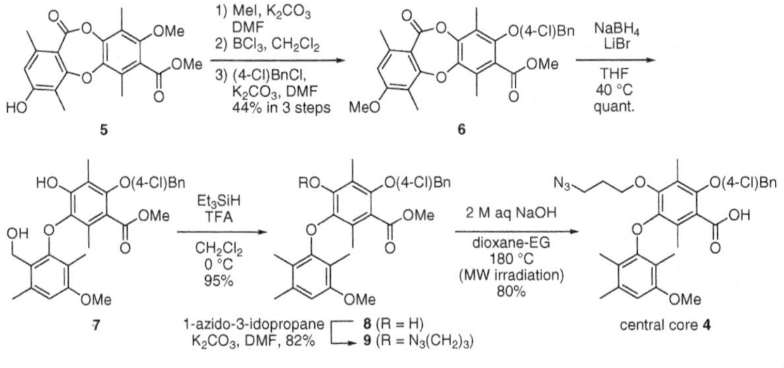

Scheme 3.2 Modified synthesis of central core **4**

3-iodopropane [3] to give **9**. Alkaline hydrolysis under microwave irradiation conditions [4] proceeded smoothly without the loss of the azide group to give **4**. The conversion from **5** based on the conventional route shown in Chap. 2 required 8 steps to provide **9** in 17 % yield. However, the work described in this chapter provided the desired compound in only 6 steps with a significant improvement in the total yield to 34 %.

3.3.2 Elongation of the Side Wings

The strategy used for the elongation of the side wings of acid **4** is shown in Scheme 3.3. Esterification of **4** with phenol **10** in the presence of excess $(CF_3CO)_2O$ [5] at 80 °C provided **11** in 78 % yield. The formylation of **11** was subsequently examined using Cl_2CHOMe–AgOTf, but this system led to the formation of a complex mixture of products at temperatures greater than −40 °C, which were required to push the reaction to completion. It has been reported that TfOH can react with azides to give the corresponding nitrenes, and a similar reaction may have occurred in this case to induce the formation of the observed

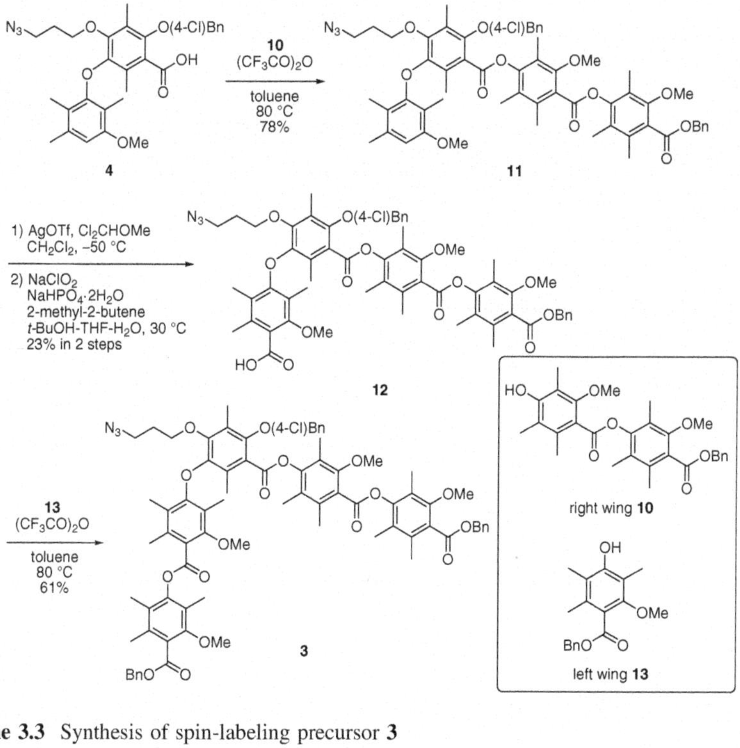

Scheme 3.3 Synthesis of spin-labeling precursor **3**

byproducts [6]. With this in mind, a solution of Cl_2CHOMe in CH_2Cl_2 was slowly added to a solution of **11** and AgOTf in CH_2Cl_2 at −78 °C, and the resulting mixture was then warmed to −50 °C to give the desired aldehyde, although it should be mentioned that the formylation did not proceed to completion. Unfortunately, compound **11** and the resulting aldehyde were difficult to separate using conventional purification techniques, and the crude product mixture was therefore subjected to a Kraus oxidation to afford acid **12**. The addition of THF to this reaction was found to be essential to dissolve the mixture and allow for the reaction to proceed to completion. The esterification of acid **12** with phenol **13** proceeded successfully with excess $(CF_3CO)_2O$ at 80 °C to furnish **3** in moderate yield. Taken together, these results show that developed synthetic route for the construction of **1** could also be useful for the synthesis of different derivatives of **1**.

3.4 Synthesis of Spin-Labeled Thielocin B1

3.4.1 Examination of Spin-Labeling by Amidation

The spin-labeling of precursor **3** was examined in detail. The removal of all of the benzyl groups from compound **3** and the reduction of the azide was achieved in one step by a hydrogenolysis reaction. While the resulting crude product was found to be poorly soluble in MeOH, MeCN and water, compound **14** could be dissolved in a 3:1 (v/v) mixture of MeCN and water to allow for its analysis by LC-MS. However, initial attempts to effect the amidation of crude amine **14** with 4-carboxy-TEMPO using DM-TMM [7, 8] as a condensation reagent failed to provide any of the desired product **2a** (Scheme 3.4).

3.4.2 Preliminary Investigation of the Spin-Labeling Process with a Model Substrate

Given that the supply of compound **3** was severely limited by the low conversion observed for the formylation of **11**, significant efforts were directed towards the development of a new synthetic methodology for the conversion of **11–15**. Considering the difficulties associated with the handling of the highly polar compound **14**, it was envisaged that the spin-labeling could be conducted prior to the removal of the protecting groups (Scheme 3.5).

The selective reduction of the azide group in **11** was initially performed in the presence of the benzyl groups. Unfortunately, the reduction of the azide group in **11** using a Staudinger reaction [9] gave phenol **17** rather than the desired amine **16**. The failure of this reaction was attributed to the electron-withdrawing carbonyl group located at the *para*-position of the 3-aminopropyloxy group in **16**, which

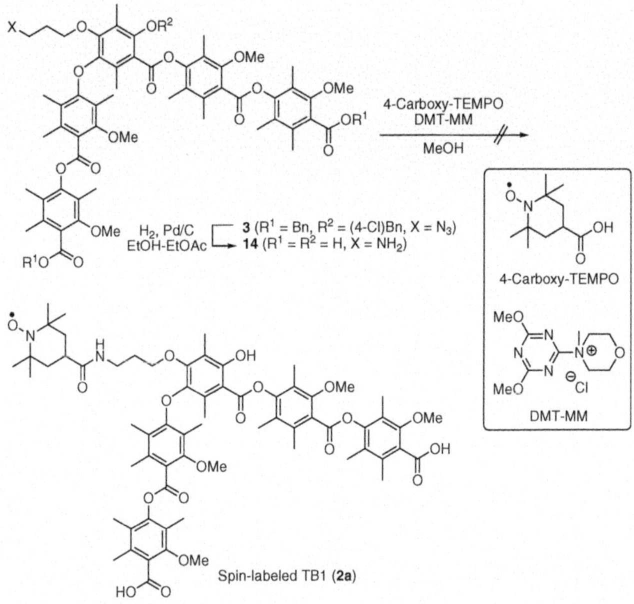

Scheme 3.4 Introduction of spin-labeling reagent with amine **13**

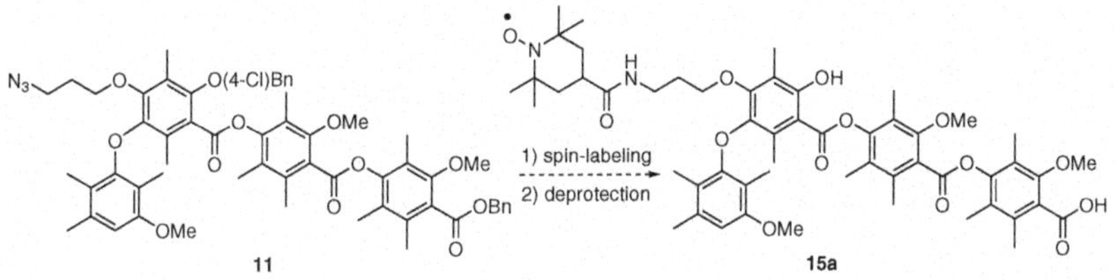

Scheme 3.5 Strategy for the synthesis of **15a**

would have induced the intramolecular cyclization of the amine product to give the corresponding azetidine together with **17** under the high-temperature conditions (Scheme 3.6). The direct amidation of azide **11** with 4-carboxy-TEMPO was subsequently examined as an alternative strategy for the introduction of the spin-labeled group. Following an extensive period of investigation, it was found that a Staudinger ligation using 2,2-dipyridyl disulfide and PMe_3 [10] afforded the desired amide **18** in moderate yield. However, subsequent hydrogenolysis to remove the benzyl groups resulted in the reduction of nitroxyl radical to provide **19** (Scheme 3.7).

Considering these results, MOM ether and ester groups were selected as protecting groups for the hydroxyl and carboxyl groups, respectively. It was envisaged that the MOM groups could be readily cleaved under acidic conditions without losing the N–O bond of the spin-labeled group. The initial removal of the benzyl

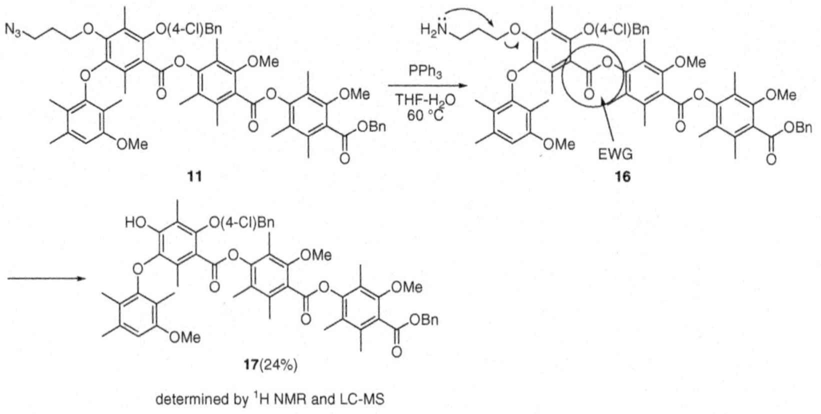

Scheme 3.6 Side reaction with the free amine in **16**

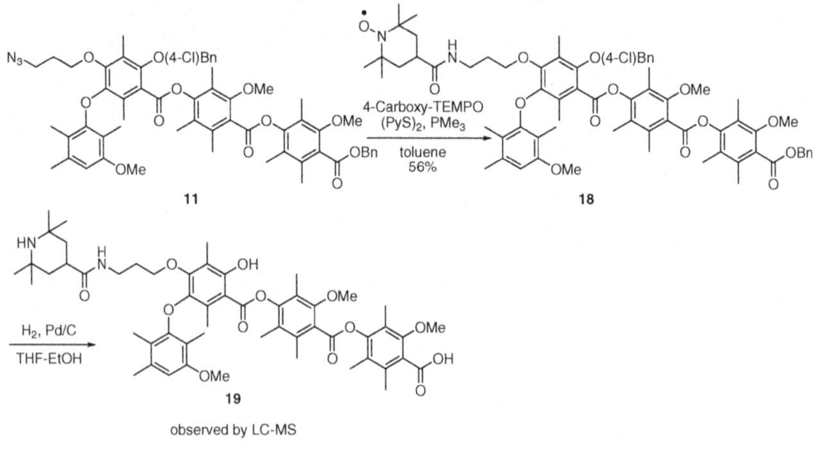

Scheme 3.7 Direct amidation of **11** by Staudinger ligation

groups in the presence of the azide group was examined in a variety of different conditions, as summarized in Table 3.1. Treatment of **11** with DDQ [11] led to the oxidation of a single methyl group to provide the undesired aldehyde **21** (Table 3.1, entry 1). The oxidation of **9** with DDQ followed by the reduction of resulting aldehyde gave benzylic alcohol **24**, which was in good agreement with the product derived from **7** by etherification, therefore confirming that the correct methyl group had been oxidized (Scheme 3.8). Treatment of **11** with BBr_3 resulted in the removal of the methyl and benzyl ether groups at the positions *ortho* to the carbonyl groups without attacking the benzyl ester (Table 3.1, entry 2). Nucleophilic cleavage of the methyl and benzyl groups with potassium trimethylsilanolate [12] in a Krapcho fashion resulted in no reaction, with **11** being recovered quantitatively (Table 3.1,

Table 3.1 Removal of the benzyl groups in **11**

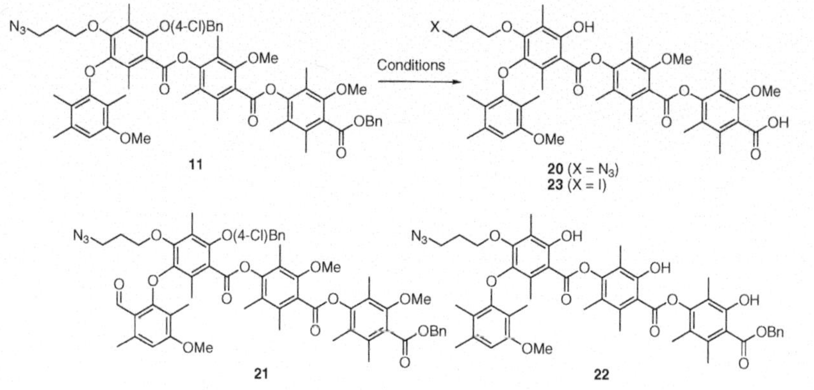

Entry	Reagent	Solv.	Conditions	Product (Yield, %)
1	DDQ	CH_2Cl_2–H_2O	30 °C, 16 h	**21** (32)
2	BBr_3	CH_2Cl_2	0 °C, 21 h	**22** (14)
3	KOTMS	THF	40 °C, 16 h	N.R.
4	Me_3SiI	CH_2Cl_2	Rt, 4 h	**23** (57)

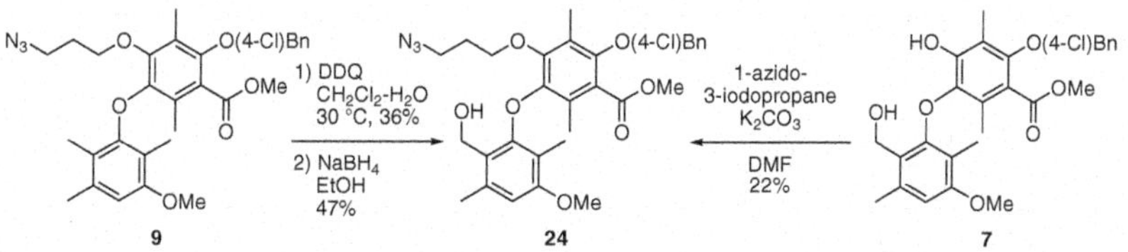

Scheme 3.8 Synthesis of benzylic alcohol **24**

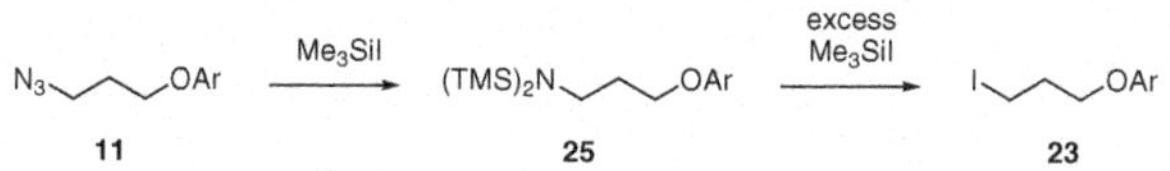

Scheme 3.9 Plausible mechanism of the substitution reaction in **11**

entry 3). In contrast, treatment of **11** with Me_3SiI [13] proved to be quite effective for the deprotection, although the azide group was substituted with iodine (Table 3.1, entry 4). The formation of the iodine species was attributed to the presence of excess Me_3SiI, which may have induced the nucleophilic substitution reaction after the reduction to silanamine **25** [14] (Scheme 3.9).

Given that the commercially available reagent can be readily deactivated, fresh Me_3SiI was prepared in situ from $(Me_3Si)_2$ and I_2 [15] and immediately used for the

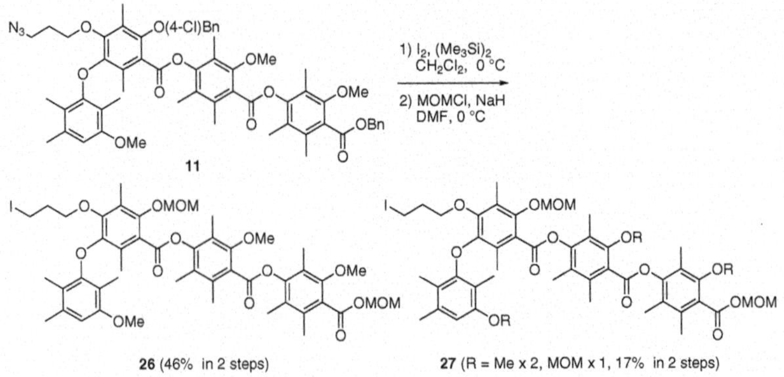

Scheme 3.10 Scalable synthesis of **23**

reaction. While the partial cleavage of one of the methyl ethers was also observed in this case, the desired deprotected product was subsequently reprotected with MOMCl to give **26**, which was purified by preparative TLC (Scheme 3.10).

In light of the unexpected substitution of the azide group with an iodide group, several attempts were made to affect the spin-labeling of **26** by *O*-alkylation, and the results of these experiments are summarized in Table 3.2. The etherification of

Table 3.2 Spin-labeling of iodine **27** by *O*-alkylation

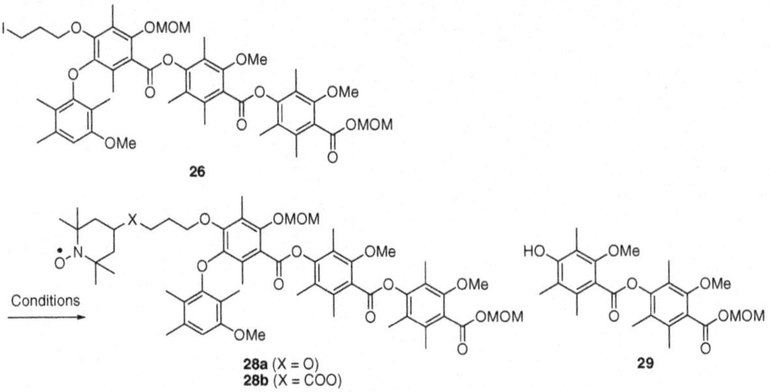

Entry	Reagent	Solv.	Conditions	Product (Yield, %)
1	4-hydroxy-TEMPO, NaH	DMF	40 °C, 2.5 h	**29** (50)
2	4-hydroxy-TEMPO, Ag_2O	MeCN	60 °C, 24 h	N.R.
3	4-hydroxy-TEMPO, Ag_2O	MeCN	120 °C (MW) 40 min	Decomp.
4	4-carboxy-TEMPO, CsF	DMF	Rt, 13 h	**28b** (79)

26 with 4-hydroxy-TEMPO was initially investigated. The in situ generation of the alkoxide of 4-hydroxy-TEMPO resulted in the cleavage of the ester bond to provide **29** (Table 3.2, entry 1). Attempts to affect the alkylation using Ag_2O under neutral conditions were also fruitless (Table 3.2, entries 2 and 3). Pleasingly, however, the carboxylate prepared from 4-carboxy-TEMPO and CsF [16] was particularly effective and reacted with **26** to afford spin-labeled **28b** in 79 % yield.

Finally, the removal of all of the MOM groups was performed according to the procedure shown in Scheme 3.11. Although the deprotection proceeded smoothly under acidic conditions, hydroxyl amine **30** was obtained as the major product. The treatment of **28b** with TFA would initially result in the formation of the oxoammonium cation species **31** [17], which would be reduced by the methanol added for the analysis of the crude product to give **30** (Scheme 3.12). Given that the use of milder acids such as AcOH did not allow for the complete cleavage of the MOM groups, we proceeded to evaluate the oxidation of **30**. After investigating several oxidants with simple handling requirements, PbO_2 [18] was found to be an efficient oxidant and gave **15b** as the major product.

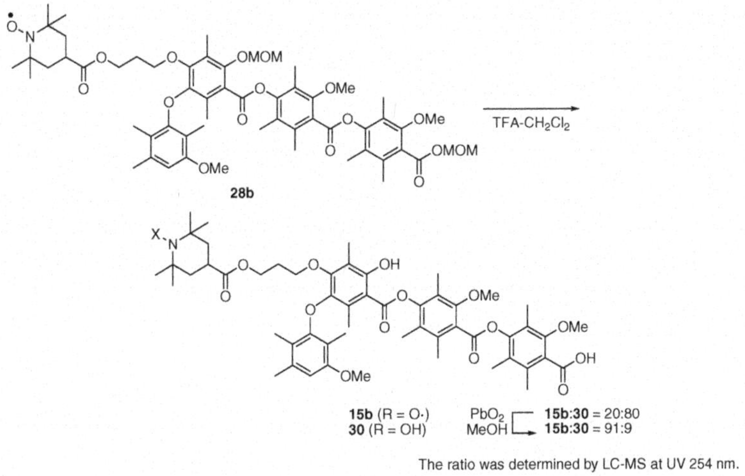

Scheme 3.11 Synthesis of **15b**

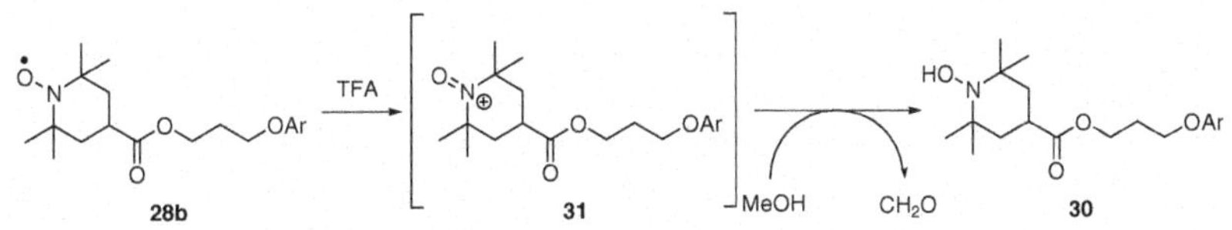

Scheme 3.12 Plausible mechanism of the reduction

3.4.3 Spin-Labeling for the Synthesis of Spin-Labeled Thielocin B1

The optimized route for the synthesis of spin-labeled thielocin B1 (**2b**) is shown in Scheme 3.13. The treatment of **3** with Et_3SiI, which was prepared in situ from I_2 and $(Me_3Si)_2$, resulted in the removal of all of the benzyl groups and the substitution of the azide group with an iodide group. Notably, no overreaction (i.e., demethylation) was observed when the reaction was conducted at −40 °C, and the resulting hydroxyl and carboxyl groups were immediately protected with MOMCl

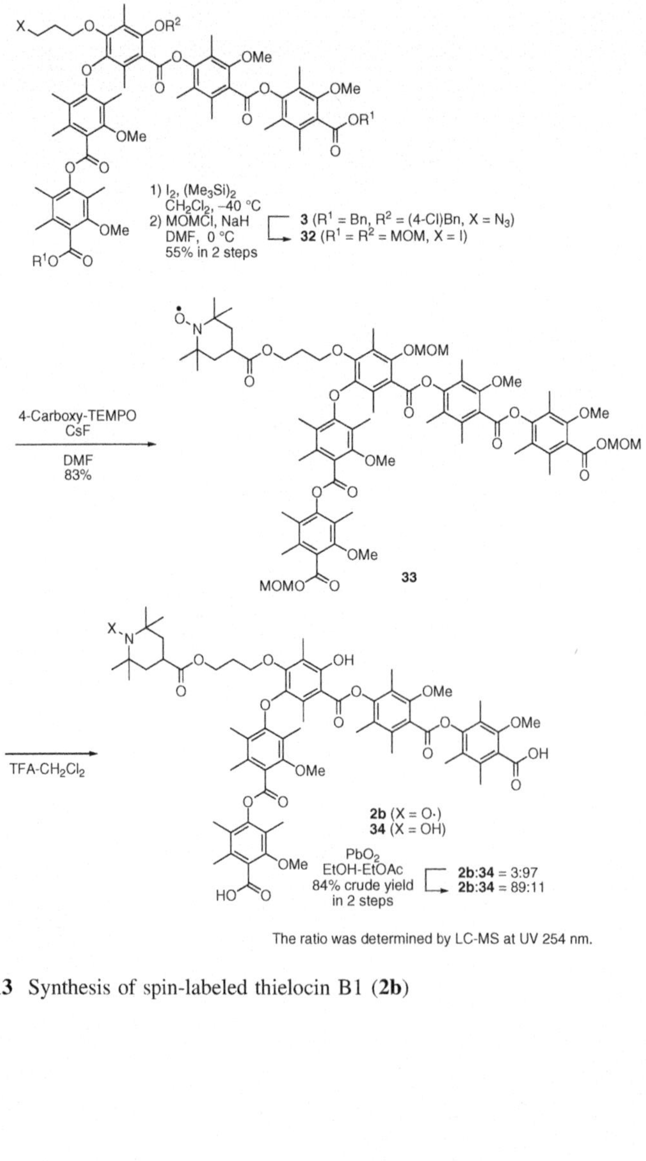

1) I_2, $(Me_3Si)_2$ CH_2Cl_2, −40 °C
2) MOMCl, NaH DMF, 0 °C
55% in 2 steps

3 (R^1 = Bn, R^2 = (4-Cl)Bn, X = N_3)
32 (R^1 = R^2 = MOM, X = I)

4-Carboxy-TEMPO CsF DMF 83%

33

TFA-CH_2Cl_2

2b (X = O·)
34 (X = OH)

PbO_2 EtOH-EtOAc 84% crude yield in 2 steps

2b:**34** = 3:97
2b:**34** = 89:11

The ratio was determined by LC-MS at UV 254 nm.

Scheme 3.13 Synthesis of spin-labeled thielocin B1 (**2b**)

to afford **32** in 55 % yield over the two steps. The *O*-alkylation of 4-carboxy-TEMPO with iodide **32** was performed in the presence of CsF to furnish **33**. Finally, the removal of all of the MOM groups followed by the reoxidation of the resulting hydroxyl amine **34** afforded the desired spin-labeled compound **2b**. Although several attempts were made to isolate **2b** by reversed-phase HPLC, they partially resulted in the reduction of the nitroxyl radical. Compounds **34** and **2b** could be separated by reversed-phase HPLC, and eluted with retention times of 5.5 and 6.2 min, respectively. The analysis of each peak by LC-MS revealed that the former was **34**, which existed as a single product. However, the latter of these two peaks was found to be a mixture of **2b** and **34** (**2b**:**34** = 69:31–78:22). These results suggested that **2b** could be undergoing a reduction during its purification over silica gel or its concentration in vacuo. Given that the contamination of the product with **34** would not pose a serious problem to the spin-labeled experiments, it was decided to use the 89:11 mixture of **2b** and **34** without separation.

3.5 Observation of the PRE Effects of Spin-Labeled Thielocin B1 on PAC3 Homodimer

Spin-labeling experiments involving PAC3 homodimer were carried out in the presence of synthesized **2b** to observe its PRE effects. A solution of **2b** in methanol-d_4 (10 mM) was titrated to ^{15}N-labeled PAC3 (0.3 mM for the monomer) to give final **2b** concentrations of 0.3, 0.6 and 0.9 mM. Samples A and B were also prepared to measure the ^{1}H–^{15}N HSQC spectra of PAC3 homodimer both in the presence and absence of the nitroxyl radical **2b**. A PBS solution of ascorbic acid (10 equivalents for **2b**), which was used as a radical quencher, or PBS alone, which was used as a reference, was added to a mixture of **2b** and PAC3 homodimer (Fig. 3.4). The intensities of the NMR signals of all of the amino acid residues on

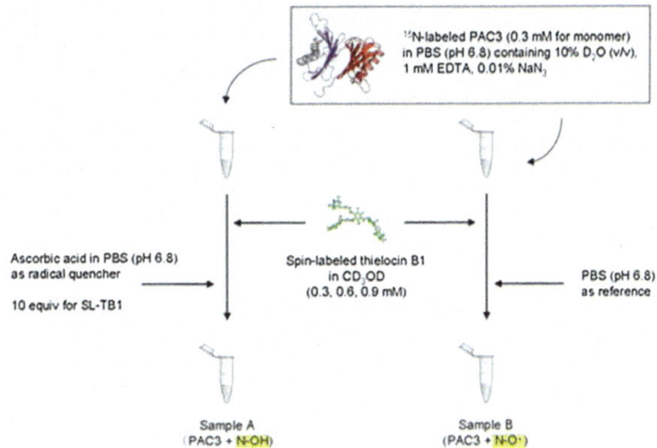

Fig. 3.4 Preparation of samples for the spin-labeling experiment. Ascorbic acid was used as a radical quencher

PAC3 were compared between samples A and B to determine the occurrence of a PRE effect. When the ratio of PAC3 (for the monomer) and **2b** was 1:1 or 1:2, the intensities of almost all of the NMR signals belonging to the NH peaks on PAC3 were decreased to an equal extent. This result indicated that most of the structure of **2b** existed independently of PAC3 homodimer. In contrast, distinct decreases were observed in the intensities of the NMR signals of PAC3 homodimer in the presence of three equivalents of **2b**. This result therefore indicated that **2b** was interacting with PAC3 in some way, and the average and standard deviation values of the NH peaks of PAC3 were consequently calculated for samples A and B after each measurement had been conducted in triplicate. These data revealed significant decreases in the intensities of the NMR signal of 16 amino acid residues on PAC3, including Thr4, Thr13, Ser29, Phe39, Lys41, Glu49, Gln70, Ile75, Val77, Val99, Lys102, Asp103, Lys104, Gly108, Lys110 and Val115 (Fig. 3.5).

The mapping of these amino acid residues onto the X-ray crystal structure of human monomeric PAC3 (PDB code: 2Z5E) [19] is shown in Fig. 3.6. Given that these residues are located in close proximity to each other on the surface of PAC3, it is clear that the nitroxyl radical in **2b** can influence the intensities of the NMR signals in the **2b**/PAC3 complex. As expected, distinct decreases were observed in the intensities of the NH peaks of PAC3 around area A, which was set as the back of the binding site with **1**. These data therefore support the suggestion that **2b** approaches PAC3 homodimer rather than monomeric PAC3.

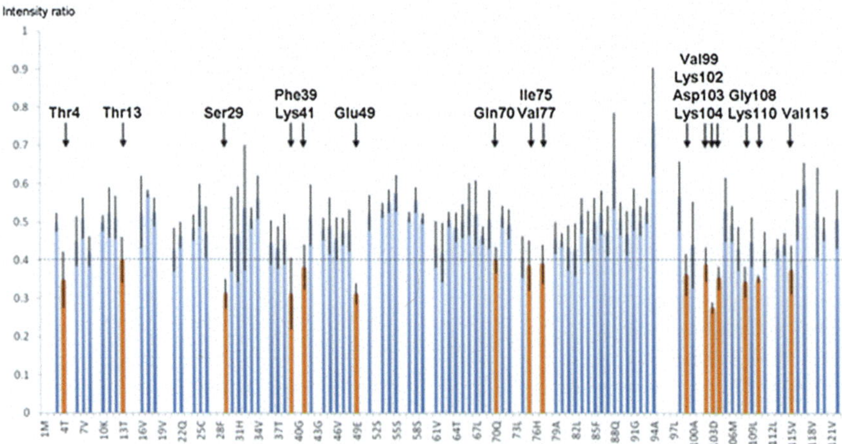

Fig. 3.5 The ratio of NMR signal intensities in the residues of PAC3 based on the results of the spin-labeling experiment with **2b**. The data shown are the mean ± S.D. Significant decreases in the peak intensities (<0.4) have been labeled

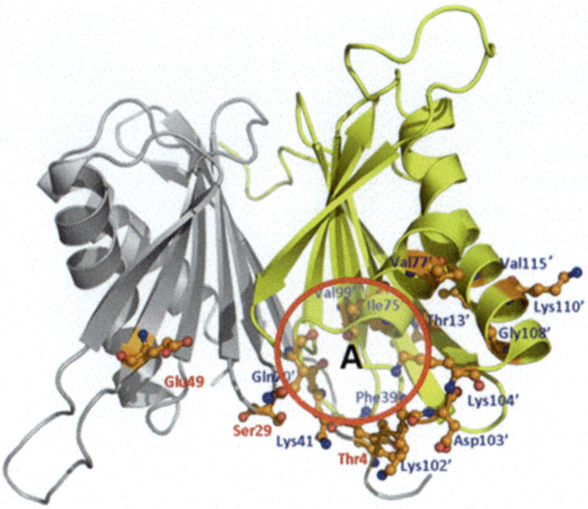

Fig. 3.6 Mapping of the residues of PAC3 homodimer exhibiting significant decreases in their peak intensities (PDB code: 2Z5E). Thr4, Thr13, Ser29, Phe39, Lys41, Glu49, Gln70, Ile75, Val77, Val99, Lys102, Asp103, Lys104, Gly108, Lys110 and Val115 are shown as carbons in *orange* (ball and stick model)

3.6 Additional Docking Study of Spin-Labeled Thielocin B1 and PAC3 Homodimer

Based on the results of the NMR experiments described in Sects. 2.7 and 3.5, an additional in silico docking study was performed involving the docking of **2b** onto the surface of PAC3 homodimer using the Molecular Operating Environment program [20]. The most significant pose obtained by this study is shown in Fig. 3.7a. This particular pose suggested that **2b** approaches from the bottom face of PAC3 homodimer. In this way, the left wing of **2b** would actually interact with the Glu49 residue and sit in close proximity to the Gly40′, Met42′ and Gly43′ residue located on the interface of PAC3 homodimer because the radical moiety in **2b** should be close to the bottom right-hand side of PAC3 homodimer based on the results of the spin-labeling experiment. Following the proposed approach of **1** to PAC3 homodimer, it is envisaged that the right wing of **1** could move towards the β-sheet layers of PAC3 homodimer to induce the dissociation to monomeric PAC3 (Fig. 3.7b).

3.7 Summary

In conclusion, the synthesis of spin-labeled thielocin B1 (**2b**) has been achieved for the first time. The spin-labeled material provided detailed information pertaining to the binding site of thielocin B1 (**1**) on the surface of PAC3 homodimer simply by

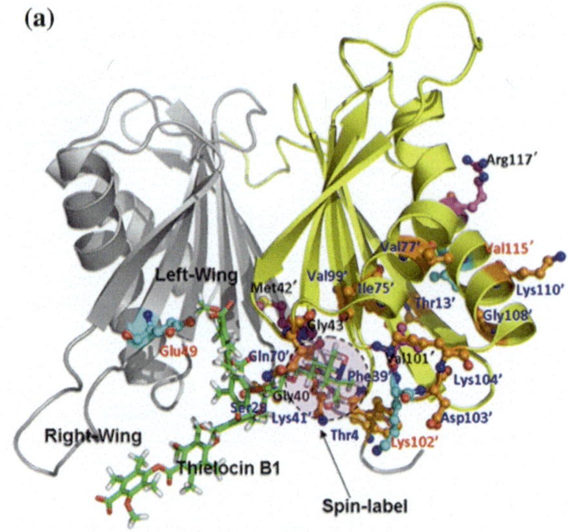

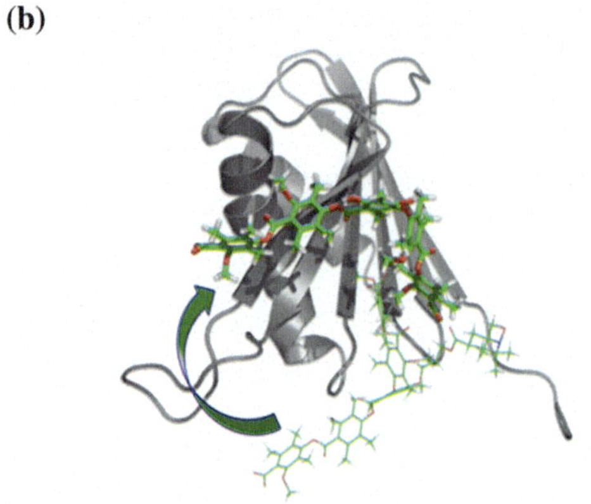

Fig. 3.7 **a** Docking model of **2b** (stick model, carbons in *green*)/PAC3 homodimer (PDB code: 2Z5E). Residues exhibiting significant changes in their chemical shifts by NMR titration with **1** are shown as carbons in *magenta* (ball and stick model). Residues exhibiting significant PRE effects with **2b** are shown as carbons in *orange*. Residues exhibiting both effects are shown as carbons in *cyan*. **b** Overlay of the MD simulation model of the complex of **1** (stick model)/PAC3 and **2b** (wire model) described in the docking model (**a**)

adding a spin-labeling reagent to the wing to the left-hand side of **1**. The PRE effect of **2b** on PAC3 led to distinct decreases in the intensities of the NMR signals of 16 residues on the surface of PAC3 homodimer, which were measured by ^{1}H–^{15}N HSQC analysis both before and after quenching the radical moiety. The results of an additional in silico docking study based on the NMR experiments suggested that the left wing of **1** formed an initial interaction with the interface of PAC3 homodimer, and that the right wing of **1** could then move towards the β-sheet layers of PAC3 to induce the dissociation of PAC3 homodimer.

3.8 Experimental Section

3.8.1 General Techniques

All commercially available reagents were used as received. Dry THF and CH_2Cl_2 (Kanto Chemical Co.) were obtained by passing commercially available pre-dried, oxygen-free formulations.

Microwave irradiation was performed with Biotage Initiator^{TM}.

All reactions were monitored by thin-layer chromatography carried out on 0.2 mm E. Merck silica gel plates (60F-254) with UV light, and visualized by *p*-anisaldehyde H_2SO_4-ethanol solution. Column chromatography and flash column chromatography were carried out with silica gel 60 N (Kanto Chemical Co. 100–210 μm) and silica gel 60 N (Kanto Chemical Co. 40–50 μm), respectively.

^{1}H NMR spectra (400 and 600 MHz) and ^{13}C NMR spectra (100 and 150 MHz) were recorded on JEOL JNM-AL400 and JEOL ECA-600 spectrometers, respectively, in the indicated solvent. Chemical shifts (δ) are reported in units parts per million (ppm) relative to the signal for internal tetramethylsilane (0.00 ppm for ^{1}H) for solutions in chloroform-*d*. NMR spectral data are reported as follows: chloroform-*d* (77.0 ppm for ^{13}C), methanol-d_4 (3.30 ppm for ^{1}H and 49.0 ppm for ^{13}C) when internal standard is not indicated. Multiplicities are reported by the following abbreviations: s (singlet), d (doublet), t (triplet), quin (quintet), m (multiplet), brs (broad singlet), J (coupling constants in Hertz).

Mass spectra and high-resolution mass spectra were measured on JEOL JMS-DX303 (for EI), SYNAPT G2 HDMS instruments (for ESI) and ThermoScientific^{TM} Exactive^{TM} Plus Orbitap Mass Spectrometer (for ESI). IR spectra were recorded on a JASCO FTIR-8400. Only the strongest and/or structurally important absorption are reported as the IR data afforded in cm^{-1}. Melting points were measured on Round Science RFS-10, and are uncorrected.

The ratio of nitroxyl radical **2b** and hydroxylamine **35** was determined by reversed-phase HPLC (Waters LC/MS system ZQ-2000) at UV 254 nm. The column was used XBridge^{TM} C18-3.5 μm, 4.6 × 7.5 mm (Waters). Gradient method is as follows: 10 % of B (0.0 min), 10–95 % of B (0.0–4.0 min), 95 % of B (4.0–11.0 min), 95–10 % of B (11.0–15.0 min), 10 % of B (15.0 min) (A: 0.1 % HCOOH/H_2O, B: 0.1 % HCOOH/MeOH).

3.8.2 Experimental Methods

Methyl 8-(4-chlorobenzyloxy)-3-methoxy-1,4,6,9-tetramethyl-11-oxo-11*H*-dibenzo[*b,e*] [1, 4] -dioxepine-7-carboxylate (6).
To a solution of phenol **5** (3.19 g, 8.57 mmol) in DMF (30 mL) were added K_2CO_3 (5.09 g, 36.9 mmol, 4.3 equiv) and MeI (1.07 mL, 17.1 mmol, 2.0 equiv) at room temperature under an argon atmosphere. After being stirred at the same temperature

for 16 h, the reaction mixture was filtered through a pad of Celite®. The filtrate was diluted with EtOAc, and acidified with 3 M aq HCl. The organic layer was separated, and the aqueous layer was extracted with EtOAc twice. The combined organic layers were washed with brine twice, saturated aq $NaHCO_3$ and brine, dried with $MgSO_4$, and filtered. The filtrate was concentrated in vacuo to give methyl ether as a white solid. The crude methyl ether was used for next reaction without further purification.

To a solution of crude methyl ether in dry CH_2Cl_2 (30 mL) was added BCl_3 (1.0 M in CH_2Cl_2, 34.3 mL, 34.3 mmol, 4.0 equiv) at 0 °C under an argon atmosphere. After being stirred at room temperature for 2.5 h, the reaction mixture was quenched with water at 0 °C. The organic layer was separated, and the aqueous layer was extracted with EtOAc twice. The combined organic layers were washed with saturated aq $NaHCO_3$ and brine, dried with $MgSO_4$, and filtered. The filtrate was concentrated in vacuo to give phenol as a white solid. The crude phenol was used for next reaction without further purification.

To a solution of the crude phenol in DMF (30 mL) were added K_2CO_3 (3.55 g, 25.7 mmol, 3.0 equiv) and (4-Cl)BnCl (1.93 g, 12.0 mmol, 1.4 equiv), at room temperature under an argon atmosphere. After being stirred at the same temperature, the reaction mixture was filtered through a pad of Celite®. The filtrate was diluted with EtOAc, and acidified with 3 M aq HCl. The organic layer was separated, and the aqueous layer was extracted with EtOAc twice. The combined organic layers were washed with brine twice, saturated aq $NaHCO_3$ and brine, dried with $MgSO_4$, and filtered. The filtrate was concentrated in vacuo, and the resulting residue was recrystallized from CH_2Cl_2/hexane to afford benzyl ether **6** (1.88 g, 3.78 mmol, 44 % in 3 steps) as a white solid. mp 159–160 °C; ^{1}H NMR (400 MHz, $CDCl_3$) δ 7.36 (2H, d, J = 8.8 Hz), 7.33 (2H, d, J = 8.8 Hz), 6.58 (1H, s), 4.82 (2H, s), 3.87 (3H, s), 3.81 (3H, s), 2.51 (3H, s), 2.38 (3H, s), 2.32 (3H, s), 2.29 (3H, s); ^{13}C NMR (100 MHz, $CDCl_3$) δ 167.7, 163.0, 161.3, 160.3, 151.0, 146.5, 144.8, 142.5, 135.2, 134.0, 129.1, 128.7, 126.4, 125.7, 122.2, 115.1, 113.6, 110.0, 75.7, 55.9, 52.4, 21.6, 14.4, 10.2, 9.9; IR (Neat) 2,950, 1,732, 1,605, 1,566, 4,582, 1,139 cm^{-1}; HREIMS calcd for $C_{27}H_{25}ClO_7$ 496.1289, found 496.1288.

Methyl 2-(4-chlorobenzyloxy)-4-hydroxy-5-(2-hydroxymethyl-5-methoxy-3,6-dimethyl-phenoxy)-3,6-dimethylbenzoate (7).

To a suspension of $NaBH_4$ (764 mg, 20.2 mmol, 20 equiv) in dry THF (30 mL) was added LiBr (1.75 g, 20.2 mmol, 20 equiv) under argon atmosphere at room temperature. After being stirred at reflux for 2 h, the reaction mixture was cooled at room temperature. To the mixture was added depsidone **6** (500 mg, 1.01 mmol) at 0 °C. After being stirred at 40 °C for 15 h, the reaction mixture was diluted with toluene, and quenched with 3 M aq HCl at 0 °C. The organic layer was separated, and the aqueous layer was extracted with EtOAc twice. The combined organic layers were washed with saturated aq $NaHCO_3$ and brine, dried with $MgSO_4$, and filtered. The filtrate was concentrated in vacuo, and the resulting residue was purified by column chromatography on silica gel (eluted with hexane/EtOAc = 2:1) to afford benzyl ether **7** (506 mg, 1.01 mmol, quant) as a white solid. mp 139–140 °C; ^{1}H NMR

(400 MHz, $CDCl_3$) δ 7.34 (4H, s), 6.49 (1H, s), 5.03 (1H, d, J = 10.8 Hz), 4.85 (1H, d, J = 11.0 Hz), 4.81 (1H, d, J = 11.0 Hz), 4.70 (1H, d, J = 10.8 Hz), 3.79 (6H, s), 2.41 (3H, s), 2.15 (3H, s), 2.10 (3H, s), 1.74 (3H, s); ^{13}C NMR (100 MHz, $CDCl_3$) δ 168.5, 158.5, 155.2, 150.3, 147.6, 140.9, 136.0, 135.8, 133.7, 129.2, 128.6, 124.5, 121.4, 120.8, 118.8, 114.7, 108.2, 75.5, 57.4, 55.6, 52.1, 19.4, 13.8, 9.4, 8.7; IR (Neat) 3,394, 2,951, 1,728, 1,611, 1,464, 1,288, 1,218, 1,123, 755 cm^{-1}; HRESI-TOFMS calcd for $C_{27}H_{30}ClO_7$ $[M+H]^+$ 501.1680, found 501.1680.

Methyl 4-benzyloxy-2-(4-chlorobenzyloxy)-5-(3-methoxy-2,5,6-trimethyl-phenoxy)-3,6-dimethylbenzoate (8).

To a solution of benzyl alcohol **7** (650 mg, 1.30 mmol) in dry CH_2Cl_2 (6.0 mL) were added TFA (0.6 mL) and Et_3SiH (837 μL, 5.19 mmol, 4.0 equiv) at 0 °C under an argon atmosphere. After being stirred at the same temperature for 30 min, the reaction mixture was diluted with EtOAc, and quenched with saturated aq $NaHCO_3$. The organic layer was separated, and the aqueous layer was extracted with EtOAc twice. The combined organic layers were washed with brine, dried with $MgSO_4$, and filtered. The filtrate was concentrated in vacuo, and the resulting residue was purified by column chromatography on silica gel (eluted with hexane/EtOAc = 15:1) to afford pentamethylbenzene **8** (601 mg, 1.24 mmol, 95 %) as a yellowish oil. ^{1}H NMR (400 MHz, $CDCl_3$) δ 7.35 (4H, s), 6.53 (1H, s), 5.97 (1H, s), 4.85 (2H, s), 3.80 (3H, s), 3.75 (1H, s), 2.26 (3H, s), 2.18 (3H, s), 2.01 (3H, s), 1.93 (3H, s), 1.80 (3H, s); ^{13}C NMR (100 MHz, $CDCl_3$) δ 168.6, 156.0, 153.4, 149.8, 147.0, 139.6, 135.8, 135.7, 133.7, 129.1, 128.6, 122.4, 121.4, 119.7, 116.6, 115.0, 108.5, 75.5, 55.7, 52.1, 20.5, 13.5, 12.3, 9.3, 9.1; IR (Neat) 3,484, 2,950, 1,726, 1,612, 1,465, 1,287, 1,205, 1,125 cm^{-1}; HREIMS calcd for $C_{27}H_{29}ClO_6$ 484.1653, found 484.1635.

Methyl 4-(3-azidopropyloxy)-2-(4-chlorobenzyloxy)-5-(3-methoxy-2,5,6-trimethylphenoxy)-3,6-dimethylbenzoate (9).

To a solution of phenol **8** (600 mg, 1.24 mmol) in DMF (4.0 mL) were added K_2CO_3 (514 mg, 3.72 mmol, 3.0 equiv) and 1-azido-3-iodopropane [3] (313 mg, 1.48 mmol, 1.2 equiv) in DMF (2.0 mL) at room temperature under an argon atmosphere. After being stirred at the same temperature for 21 h, the reaction mixture was diluted with EtOAc, and quenched with 3 M aq HCl. The organic layer was separated, and the aqueous layer was extracted with ethyl acetate twice. The combined organic layers were washed with brine twice, saturated aq $NaHCO_3$ and brine, dried with Na_2SO_4, and filtered. The filtrate was concentrated in vacuo, and the resulting residue was purified by column chromatography on silica gel (eluted with hexane/EtOAc = 15:1) to afford 3-azidopropyl ether **9** (581 mg, 1.02 mmol, 82 %) as a colorless oil. ^{1}H NMR (400 MHz, $CDCl_3$) δ 7.35 (4H, s), 6.49 (1H, s), 4.85 (2H, s), 3.81 (3H, s), 3.80 (3H, s), 3.48-3.59 (2H, m), 3.17 (2H, t, J = 6.6 Hz), 2.26 (3H, s), 2.13 (3H, s), 2.11 (3H, s), 2.00 (3H, s), 1.91 (3H, s), 1.54 (2H, quin, J = 6.6 Hz); ^{13}C NMR (100 MHz, $CDCl_3$) δ 168.5, 155.8, 154.7, 148.9, 148.5, 145.8, 135.7, 134.8, 133.8, 129.1, 128.6, 125.9, 124.4, 124.0, 119.5, 114.5, 107.5, 75.4, 69.4, 55.8, 52.2, 48.1, 28.9, 20.4, 13.8, 12.5, 10.0, 9.4; IR (Neat) 2,925, 2,097, 1,731, 1,613, 1,283, 1,125 cm^{-1}; HREIMS calcd for $C_{30}H_{34}ClN_3O_6$ 567.2136, found 574.2108.

4-(3-Azidopropyloxy)-2-(4-chlorobenzyloxy)-5-(3-methoxy-2,5,6-trimethylphenoxy)-3,6-dimethylbenzoic acid (4).
To a solution of methyl ester **9** (580 mg, 1.02 mmol) in dioxane (5.0 mL) were added 2 M aq NaOH (10 mL, 20.0 mmol, 20 equiv) and ethylene glycol (2.5 mL) at room temperature. After being stirred at 180 °C under microwave irradiation for 50 min, the reaction mixture was cooled to room temperature, diluted with EtOAc, and quenched with 3 M aq HCl. The organic layer was separated, and the aqueous layer was extracted with EtOAc twice. The combined organic layers were washed with brine, dried with Na_2SO_4, and filtered. The filtrate was concentrated in vacuo, and the resulting residue was purified by column chromatography on silica gel (eluted with hexane/EtOAc = 4:1) to afford carboxylic acid **4** (455 mg, 821 μmol, 80 %) as an orange oil. ^{1}H NMR (400 MHz, $CDCl_3$) δ 7.36 (2H, d, J = 8.0 Hz), 7.32 (2H, d, J = 8.0 Hz), 6.50 (1H, s), 4.87 (2H, s), 3.80 (3H, s), 3.49-3.60 (2H, m), 3.18 (2H, t, J = 6.4 Hz), 2.27 (3H, s), 2.25 (3H, s), 2.15 (3H, s), 2.01 (3H, s), 1.91 (3H, s), 1.55 (2H, quin, J = 6.4 Hz); ^{13}C NMR (100 MHz, $CDCl_3$) δ 170.7, 155.9, 154.7, 149.3, 149.2, 146.1, 135.1, 134.9, 134.1, 129.4, 128.7, 125.7, 124.1, 119.5, 114.4, 107.6, 75.8, 69.5, 55.8, 48.1, 28.9, 20.5, 14.0, 12.6, 10.0, 9.4; IR (Neat) 3,316, 2,926, 2,097, 1,703, 1,613, 1,465, 1,370, 1,125, 1,092 cm^{-1}; HREIMS calcd for $C_{29}H_{32}ClN_3O_6$ 553.1980, found 553.1972.

Preparation of ester 11.
To a solution of carboxylic acid **4** (450 mg, 812 μmol) in dry toluene (10 mL) were added phenol **10** (480 mg, 975 μmol, 1.2 equiv) and $(CF_3CO)_2O$ (3.95 mL, 28.4 mmol, 35 equiv) at room temperature under an argon atmosphere. After being stirred at 80 °C for 13 h, the reaction mixture was cooled at room temperature, and quenched with 2 M aq NaOH. The organic layer was separated, and the aqueous layer was extracted with EtOAc twice. The combined organic layers were washed with brine, dried with Na_2SO_4, and filtered. The filtrate was concentrated in vacuo, and the resulting residue was purified by flash column chromatography on silica gel (eluted with hexane/EtOAc = 9:1) to afford ester **11** (651 mg, 633 μmol, 78 %) as a white solid. mp 92–93°C; ^{1}H NMR (400 MHz, $CDCl_3$) δ 7.46–7.47 (2H, m), 7.30–7.40 (7H, m), 6.52 (1H, s), 5.40 (2H, s), 4.94 (2H, s), 3.82 (3H, s), 3.78 (3H, s), 3.71 (3H, s), 3.53–3.61 (2H, m), 3.19 (2H, t, J = 6.8 Hz), 2.39 (3H, s), 2.35 (3H, s), 2.29 (3H, s), 2.25 (3H, s), 2.21 (6H, s), 2.18 (3H, s), 2.14 (3H, s), 2.10 (3H, s), 2.05 (3H, s), 1.96 (3H, s), 1.56 (2H, quin, J = 6.8 Hz); ^{13}C NMR (100 MHz, $CDCl_3$) δ 168.3, 166.2, 165.8, 155.9, 154.7, 154.2, 153.6, 150.0, 149.6, 149.4, 149.2, 146.0, 135.6, 135.4, 134.9, 133.8, 133.4, 132.7, 129.2, 128.5, 128.3, 127.4, 126.5, 126.1, 125.7, 125.1, 124.9, 124.3, 122.4, 122.2, 119.5, 114.5, 107.6, 75.7, 69.5, 67.0, 62.1, 62.0, 55.8, 48.1, 28.9, 20.5, 17.3, 16.8, 14.5, 13.4, 13.0, 12.6, 10.6, 10.2, 9.5; IR (Neat) 2,926, 2,097, 1,735, 1,578, 1,461, 1,278, 1,150, 1,122 cm^{-1}; HRESI-TOFMS calcd for $C_{58}H_{62}ClN_3O_{12}Na$ [M+Na]$^+$ 1,050.3920, found 1,050.3896.

Preparation of carboxylic acid 12.
To a solution of ester **11** (200 mg, 194 μmol) and AgOTf (250 mg, 972 μmol, 5.0 equiv) in dry CH_2Cl_2 (12 mL) was added Cl_2CHOMe (88.0 μL, 972 μmol, 5.0 equiv) in dry CH_2Cl_2 (4.0 mL) at −78 °C under an argon atmosphere. After being

stirred at −50 °C for 10 min, the reaction mixture was quenched with saturated aq $NaHCO_3$ at 0 °C. After being stirred at room temperature for 30 min, the reaction mixture was filtered through a pad of Celite®. The organic layer of the filtrate was separated, and the aqueous layer was extracted with EtOAc twice. The combined organic layers were washed with brine, dried with Na_2SO_4, and filtered. The filtrate was concentrated in vacuo to give the mixture of desired aldehyde and ester **11** as a yellowish oil. The crude mixture was used for next reaction without further purification.

To a solution of crude mixture in *t*BuOH (2.0 mL), THF (1.0 mL) and water (3.0 mL) were added 2-methyl-2-butene (2.0 mL), $NaClO_2$ (87.7 mg, 970 μmol, 5.0 equiv) and $NaH_2PO_4{\cdot}2H_2O$ (151 mg, 970 μmol, 5.0 equiv) at 0 °C. After being stirred at 30 °C for 13 h, the organic layer was separated, and the aqueous layer was extracted with EtOAc twice. The combined organic layers were washed with brine, dried with Na_2SO_4, and filtered. The filtrate was concentrated in vacuo, and the resulting residue was purified by flash column chromatography on silica gel (eluted with $CHCl_3$/MeOH = 9:1) to afford carboxylic acid **12** (47.8 mg, 44.6 μmol, 23 % in 2 steps) as a yellowish oil. Ester **11** was recovered (69.2 mg, 67.2 μmol, 35 % in 2 steps) as a yellowish oil. ^{1}H NMR (400 MHz, $CDCl_3$) δ 7.45–7.47 (2H, m), 7.31–7.40 (7H, m), 5.40 (2H, s), 4.96 (2H, s), 3.79 (3H, s), 3.78 (3H, s), 3.71 (3H, s), 3.53–3.61 (2H, m), 3.14 (2H, t, J = 6.6 Hz), 2.40 (3H, s), 2.36 (3H, s), 2.34 (3H, s), 2.25 (3H, s), 2.22 (3H, s), 2.21 (3H, s), 2.18 (3H, s), 2.15 (3H, s), 2.14 (3H, s), 2.10 (3H, s), 2.02 (3H, s), 1.56 (2H, quin, J = 6.6 Hz); ^{13}C NMR (100 MHz, $CDCl_3$) δ 171.4, 168.3, 166.2, 165.6, 156.0, 154.3, 154.1, 153.7, 150.1, 150.0, 149.4, 149.2, 145.4, 135.7, 135.3, 133.9, 133.5, 133.3, 132.7, 129.2, 128.57, 128.55, 128.53, 128.3, 127.4, 126.6, 126.1, 125.7, 125.1, 124.5, 124.1, 122.4, 122.2, 119.7, 75.8, 69.6, 67.1, 62.2, 62.1, 62.0, 47.9, 28.9, 17.3, 16.9, 16.7, 14.5, 13.4, 13.1, 13.0, 10.6, 10.3, 10.2, 10.1; IR (Neat) 3,444, 2,931, 2,098, 1,739, 1,733, 1,601, 1,575, 1,463, 1,279, 1,150, 755 cm^{-1}; HRESITOFMS calcd for $C_{59}H_{62}N_3ClO_{14}Na$ $[M+Na]^+$ 1,094.3818, found 1,094.3801.

Preparation of ester 3.

To a solution of carboxylic acid **12** (169 mg, 158 μmol) in dry toluene (10 mL) were added phenol **13** (56.8 mg, 189 μmol, 1.2 equiv) and $(CF_3CO)_2O$ (660 μL, 4.74 mmol, 30 equiv) at room temperature under an argon atmosphere. After being stirred at 80 °C for 17 h, the reaction mixture was cooled at room temperature, and quenched with 2 M aq NaOH. The organic layer was separated, and the aqueous layer was extracted with EtOAc twice. The combined organic layers were washed with brine, dried with $MgSO_4$, and filtered. The filtrate was concentrated in vacuo, and the resulting residue was purified by flash column chromatography on silica gel (eluted with hexane/EtOAc = 5:1) to afford ester **3** (131 mg, 96.8 μmol, 61 %) as a colorless oil. ^{1}H NMR (400 MHz, $CDCl_3$) δ 7.46–7.48 (4H, m), 7.33–7.40 (10H, m), 5.40 (4H, s), 4.96 (2H, s), 3.78 (6H, s), 3.71 (3H, s), 3.70 (3H, s), 3.58–3.66 (2H, m), 3.14 (2H, dt, J = 6.4, 2.0 Hz), 2.41 (3H, s), 2.40 (3H, s), 2.36 (3H, s), 2.25 (6H, s), 2.24 (3H, s), 2.21 (6H, s), 2.18 (9H, s), 2.14 (3H, s), 2.10 (3H, s), 2.06 (3H, s), 1.59 (2H, quin, J = 6.4 Hz); ^{13}C NMR (100 MHz, $CDCl_3$) δ 168.30,

168.28, 166.3, 166.2, 165.6, 156.2, 154.5, 154.3, 153.6, 150.2, 149.9, 149.5, 149.4, 149.2, 145.4, 135.6, 135.3, 133.9, 133.5, 133.3, 132.7, 129.2, 128.57, 128.55, 128.53, 128.3, 127.40, 127.37, 126.6, 126.1, 125.7, 125.6, 125.12, 125.08, 124.5, 124.1, 123.8, 122.4, 122.2, 119.9, 75.8, 69.6, 67.07, 67.06, 62.10, 62.06, 62.0, 47.9, 28.9, 17.3, 17.2, 16.8, 14.5, 13.5, 13.2, 12.97, 12.95, 10.6, 10.34, 10.29, 10.22, 10.21; IR (Neat) 2,940, 2,097, 1,734, 1,576, 1,457, 1,279, 1,148 cm^{-1}; HRESITOFMS calcd for $C_{77}H_{80}ClN_3O_{17}Na$ $[M+Na]^+$ 1,376.5074, found 1,376.5076.

Preparation of MOM ester 32.

To a solution of I_2 (42.0 mg, 166 μmol, 15 equiv) in dry CH_2Cl_2 (2.4 mL) was added hexamethyldisilane (33.4 μL, 166 μmol, 15 equiv) at room temperature under an argon atmosphere. After being stirred at 40 °C for 30 min, the reaction mixture was cooled at room temperature. To the resulting brown solution was added benzyl ester **3** (15.0 mg, 11.1 μmol) in dry CH_2Cl_2 (0.6 mL) dropwise at −78 °C under an argon atmosphere. After being stirred at −40 °C for 9 min, the reaction mixture was quenched with 1 M aq HCl at −40 °C, and stirred for 15 min at room temperature. The mixture was concentrated in vacuo to give carboxylic acid. The crude carboxylic acid was used for next reaction without further purification.

The solution of crude carboxylic acid in DMF (1.0 mL) were added NaH (50–72 % in oil, 6.5 mg, 166 μmol, 15 equiv) and MOMCl (12.6 μL, 166 μmol, 15 equiv) at 0 °C under an argon atmosphere. After being stirred at the same temperature for 15 min, the reaction mixture was diluted with EtOAc, and quenched with 1 M aq HCl. The organic layer was separated, and the aqueous layer was extracted with EtOAc twice. The combined organic layers were washed with brine twice, saturated aq $NaHCO_3$ and brine, dried with Na_2SO_4, and filtered. The filtrate was concentrated in vacuo, and the resulting residue was purified by flash column chromatography on silica gel (eluted with hexane/EtOAc = 3:1) to afford MOM ester **32** (7.7 mg, 60.8 μmol, 55 %) as an orange oil. ^1H NMR (400 MHz, $CDCl_3$) δ 5.50 (4H, s), 5.06 (2H, s), 3.85 (3H, s), 3.82 (6H, s), 3.79 (3H, s), 3.58 (9H, s), 3.56-3.64 (2H, m), 3.10 (2H, t, $J = 6.8$ Hz), 2.42 (3H, s), 2.41 (6H, s), 2.33 (3H, s), 2.32 (3H, s), 2.28 (9H, s), 2.27 (6H, s), 2.25 (6H, s), 2.18 (3H, s), 2.05 (3H, s), 1.82 (2H, quin, $J = 6.8$ Hz, z); ^{13}C NMR (100 MHz, $CDCl_3$) δ 168.1, 168.0, 166.4, 166.2, 165.4, 156.2, 154.6, 154.4, 153.6, 150.0, 149.6, 149.5, 149.3, 148.9, 145.4, 133.6, 133.3, 132.57, 132.55, 128.5, 127.3, 127.2, 126.6, 126.1, 125.8, 125.2, 125.1, 125.0, 124.1, 123.8, 122.4, 122.2, 120.0, 101.3, 91.1, 72.4, 62.2, 62.11, 62.07, 57.9, 57.8, 33.3, 17.34, 17.29, 16.7, 14.5, 13.4, 13.2, 13.1, 13.0, 11.1, 10.7, 10.4, 10.3, 10.2; IR (Neat) 2,940, 1,743, 1,678, 1,575, 1,463, 1,323, 1,279, 1,146, 1,100, 1,077 cm^{-1}; HRESITOFMS calcd for $C_{62}H_{75}IO_{20}Na$ $[M+Na]^+$ 1,289.3794, found 1,279.3789.

Preparation of ester 33.

To a solution of iodide **32** (45.0 mg, 35.5 μmol) in DMF (2.0 mL) were added CsF (27.0 mg, 178 μmol, 5.0 equiv) and 4-carboxyl-TEMPO (13.0 mg, 64.9 μmol, 1.8 equiv) at room temperature under an argon atmosphere. After being stirred at the same temperature for 17 h, the reaction mixture was diluted with EtOAc, and quenched with saturated aq $NaHCO_3$. The organic layer was separated, and the aqueous layer was extracted with ethyl acetate twice. The combined organic layers were washed with brine, dried with Na_2SO_4, and filtered. The filtrate was concentrated in vacuo, and the resulting residue was purified by column chromatography on silica gel (eluted with hexane/EtOAc = 6:1) to afford ester **33** (39.5 mg, 29.5 μmol, 83 %) as an orange solid. mp 87–88°C; ^{1}H NMR (400 MHz, $CDCl_3$) δ 5.50 (4H, s), 5.06 (2H, s), 4.17 (2H, brs), 3.85 (3H, s), 3.82 (6H, s), 3.79 (3H, brs), 3.58 (11H, s), 2.42 (6H, s), 2.41 (3H, s), 2.32 (3H, s), 2.28 (9H, s), 2.27 (9H, s), 2.25 (6H, s), 2.20 (3H, brs), 2.07 (3H, brs), 2.05 (3H, s), 1.69 (2H, brs); ^{13}C NMR (100 MHz, $CDCl_3$) δ 166.7, 166.6, 165.0, 164.8, 164.1, 154.9, 153.2, 153.0, 152.30, 152.26, 148.6, 148.24, 148.16, 148.1, 147.6, 144.0, 132.2, 131.9, 131.3, 131.2, 126.0, 125.9, 125.3, 124.7, 124.41, 124.37, 123.8, 123.7, 123.6, 122.8, 122.5, 121.0, 120.9, 118.6, 100.2, 89.9, 89.8, 67.9, 61.1, 60.9, 60.8, 60.4, 57.1, 57.0, 56.6, 27.5, 16.0, 15.7, 15.4, 13.2, 12.1, 12.0, 11.9, 11.7, 9.4, 9.3, 9.1, 9.0, 8.9; IR (Neat) 2,940, 1,739, 1,576, 1,462, 1,279, 1,146 cm^{-1}; HRESITOFMS calcd for $C_{72}H_{92}NO_{23}Na$ [M+Na]$^+$ 1,361.5958, found 1,361.5952.

Preparation of spin-labeled thielocin B1 (2b).
To a solution of MOM ester **33** (10.0 mg, 7.5 μmol) in dry CH_2Cl_2 (1.0 mL) was added TFA (0.2 mL) at room temperature. After being stirred at the same temperature for 30 min, the reaction mixture was concentrated in vacuo. The resulting residue was solved in MeOH (1 mL), and the solution was concentrated in vacuo to give the mixture of desired nitroxyl radical **2b** and hydroxylamine **34** as a white solid. The ratio of **2b** and **34** was determined by reversed-phase HPLC (**2b**:**34** = 3:97). The crude mixture was used for next reaction without further purification.

To the solution of crude mixture in EtOH (2.0 mL) and EtOAc (1.0 mL) was added PbO_2 (8.9 mg, 37.4 μmol, 5.0 equiv) at room temperature. After being stirred at the same temperature for 25 h, the reaction mixture was filtered through a pad of Celite®. The filtrate was concentrated in vacuo, and to resulting residue was washed with $CDCl_3$. The supernatant was removed, and the resulting residue was dried in vacuo to give the mixture of desired nitroxyl radical **2b** and hydroxylamine **34** as a white solid. The ratio of **2b** and **34** was determined by reversed-phase HPLC (**2b**:**34** = 89:11). The crude mixture (7.6 mg, 84 % crude yield in 2 steps) was used for next spin-label experiment without further purification. **2b**: mp > 300°C; ^{1}H NMR (600 MHz, CD_3OD) δ 4.08 (2H, brs), 3.83 (3H, s), 3.82 (6H, s), 3.77 (3H, s), 3.63 (2H, m), 2.68 (3H, brs), 2.41 (3H, s), 2.39 (3H, brs),2.28 (6H, s), 2.25 (6H, s), 2.23 (6H, s), 2.22 (6H, s), 2.18 (6H, s), 2.05 (3H, brs), 1.69 (2H, brs); ^{13}C NMR (150 MHz, CD_3OD) δ 170.3, 168.1, 167.6, 163.2, 162.9, 158.1, 156.1, 155.7, 154.2, 154.1, 154.0, 151.0, 149.8, 149.6, 143.6, 134.7, 134.5, 132.7, 132.6, 129.3, 128.1, 127.1, 126.5, 125.1, 124.8, 123.4, 122.9, 121.6, 119.6, 112.0, 70.8, 63.0, 62.9, 62.8, 62.3, 17.8, 17.6, 17.0, 16.1, 13.54, 13.45, 13.4, 11.0, 10.8, 10.6, 9.9; IR (Neat) 3,385, 2,935, 1,685, 1,577, 1,458, 1,309, 1,208, 1,187, 1,147 cm^{-1}; HPLC retention time: 10.7 min; HRESITOFMS calcd for $C_{66}H_{81}NO_{20}$ [M+H]$^+$ 1,207.5352, found 1,207.5388. **34**: mp > 300°C; ^{1}H NMR (600 MHz, CD_3OD) δ 3.96–4.05 (2H, m), 3.83 (3H, s), 3.803 (3H, s), 3.799 (3H, s), 3.76 (3H, s),

3.66–3.69 (1H, m), 3.56–3.61 (1H, m), 2.92–2.96 (1H, m), 2.68 (3H, s), 2.42 (3H, s), 2.39 (3H, s), 2.273 (3H, s), 2.265 (3H, s), 2.26 (3H, s), 2.25 (6H, s), 2.24 (3H, s), 2.21(6H, s), 2.18 (3H, s), 2.16 (3H, s), 2.10–2.12 (2H, m), 2.05 (3H, s), 1.82 (2H, t, J = 13.8 Hz), 1.61–1.69 (2H, m), 1.384 (6H, s), 1.379 (6H, s); ^{13}C NMR (150 MHz, CD_3OD) δ 174.0, 172.1, 172.0, 170.3, 168.1, 167.6, 160.4, 158.1, 156.0, 155.8, 154.5, 154.1, 151.1, 150.5, 150.4, 143.6, 134.7, 134.5, 133.2, 129.9, 129.3, 127.9, 127.2, 126.9, 125.1, 124.7, 123.4, 123.3, 120.6, 119.7, 118.0, 116.0, 122.5, 111.0, 70.7, 69.0, 68.9, 63.2, 62.9, 62.7, 62.5, 40.0, 34.3, 29.9, 28.1, 26.5, 19.9, 17.63, 17.56, 16.9, 16.12, 16.09, 13.4, 10.7, 9.9; IR (Neat) 3,349, 2,923, 2,853, 1,739, 1,676, 1,593, 1,459, 1,413, 1,154 cm^{-1}; HPLC retention time: 9.0 min; HRESITOFMS calcd for $C_{66}H_{82}NO_{20}$ $[M+H]^+$ 1,208.5430, found 1,208.5430.

NMR experiment for PRE effects.
^{15}N-labeled PAC3 was dissolved at a concentration of 0.3 mM in PBS (pH 6.8) containing 10 % D_2O (v/v), 1 mM EDTA, and 0.01 % NaN_3 and a 10 mM methanol-d_4 solution of **2b** was added to this solution. PRE effects were measured from the peak intensity ratios between two ^{1}H–^{15}N HSQC spectra of ^{15}N-labeled PAC3 with **2b** (0.3, 0.6, or 0.9 mM) in presence and absence of L-(+)-ascorbic acid (3, 6, or 9 mM) for radical quenching. All NMR data were acquired at 303 K using a Bruker AVANCE800 spectrometer equipped with a cryogenic probe. For ^{1}H–^{15}N HSQC measurements, spectra were recorded at a ^{1}H observation frequency of 800 MHz with 256 (t_1) × 2,048 (t_2) complex points and 16 scans per t_1 increment. The data were processed utilizing Bruker TopSpin 2.1 software and analyzed with SPARKY 3 [21] software.

References

1. Brown HC, Narasimhan S, Choi YM. J Org Chem. 1982;47:4702–8.
2. West CT, Donnelly SJ, Kooistra DA, Doyle MP. J Org Chem. 1973;38:2675–81.
3. Yao L, Smith BT, Aubé J. 1-azido-3-iodopropane was prepared by reported procedure. J Org Chem. 2004;69:1720–2.
4. Kappe CO. Angew Chem Int Ed. 2004;43:6250–84.
5. Parish RC, Stock LM. J Org Chem. 1965;30:927–9.
6. Song S, Zhao Y-M, Zhai H. Org Lett. 2011;13:6331–3.
7. Kunishima M, Kawachi C, Iwasaki F, Terao K, Tani S. Tetrahedron Lett. 1999;40:5327–30.
8. Kunishima M, Kawachi C, Hioki K, Terao K, Tani S. Tetrahedron. 2001;57:1551–8.
9. Staudinger H, Meyer J. Helv Chim Acta. 1919;2:635–46.
10. Burés J, Martin F, Urpi F, Vilarrasa J. J Org Chem. 2009;74:2203–6.
11. Tanaka T, Oikawa Y, Nakajima N, Hamada T, Yoshimitsu O. Chem Pharm Bull. 1987; 35:2203–8.
12. Lovrić M, Cepanec I, Litvić M, Bartolinčić A, Vinković V. Croat Chem Acta. 2007; 80:109–15.
13. Jung ME, Lyster MA. J Org Chem. 1977;42:3761–4.
14. Kamal A, Rao NV, Laxman E. Tetrahedron Lett. 1997;38:6945–8.
15. Olah JA, Narang SC, Gupta BGB, Malhotra R. Angew Chem Int Ed. 1979;18:612–4.

16. Sato T, Otera J, Nozaki H. J Org Chem. 1992;57:2166–9.
17. Tomizawa M. Ph.D. thesis, Tohoku University, Japan; 2009.
18. Bakunov S, Denicov AY, Tkachev AV. Tetrahedron. 1995;51:8565–72.
19. Yashiroda H, Mizushima T, Okamoto K, Kameyama T, Hayashi H, Kishimoto T, Niwa S, Kasahara M, Kurimoto E, Sakata E, Takagi K, Suzuki A, Hirano Y, Murata S, Kato K, Yamane T, Tanaka K. Nat Struct Mol Biol. 2008;15:228–36.
20. Molecular Operating Environment (MOE), ver. 2012. Chemical Computing Group Inc., Quebec; 2012.
21. Goddard TD, Kneller DG. SPARKY 3. University of California, San Francisco.

Chapter 4
An Improved Method for the Direct Formylation of Substituted Benzenes Using Dichloromethyl Methyl Ether-Silver Trifluoromethanesulfonate

4.1 Introduction

As mentioned in Sect. 2.3.3, the use of dichloromethyl methyl ether (Cl_2CHOMe) in conjunction with trifluoromethanesulfonate (AgOTf) was found to be effective for the formylation of highly functionalized benzenes. Based on our success with this reaction, the decision was taken to evaluate its reactivity and functional group tolerance in comparison with several conventional methods for the formylation of substituted benzene. This chapter therefore provides a detailed description of the scope and limitations of this newly discovered formylation method.

4.2 Formylation of Substituted Phenyl Ethers

The formylation of simple *O*-protected phenols **2** was selected as a model reaction for our initial investigation of this novel formylation protocol. It is well known that it can be difficult to formylate alkoxybenzenes at their *ortho*-position because they provide higher levels of steric hindrance and are less electron-donating than the corresponding phenols. Furthermore, ether-type protecting groups on the *ortho*-phenol groups are invariably removed by chelating effects under the highly acidic conditions required by these reactions to generate undesired **3** (Fig. 4.1) [1–3].

The results for the formylation of 4-benzyloxytoluene (**2a**) in a variety of different conditions are summarized in Table 4.1. The fact that a compound as simple as **2a** cannot be directly formylated under conventional formylation conditions provides a good indication of the difficulties associated with the formylation of protected phenols [4]. In practice, Vilsmeier-Haack reactions using $POCl_3$ in DMF [5] or N-methylformanilide (NMFA) [6] failed to provide any of the desired product, with **1a** being recovered completely (Table 4.1, entries 1 and 2). A modified Duff formylation involving the treatment of **2a** with hexamethylenetetramine (HMTA) in TFA [7]

K. Ohsawa, *Total Synthesis of Thielocin B1 as a Protein–Protein Interaction Inhibitor of PAC3 Homodimer*, Springer Theses, DOI 10.1007/978-4-431-55447-9_4

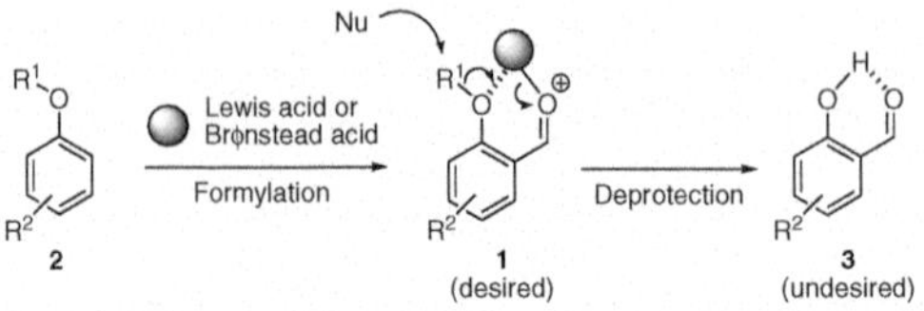

Fig. 4.1 Side reaction of synthesized **1** under acidic conditions

Table 4.1 Optimization of the reaction conditions for the formylation of **2a**

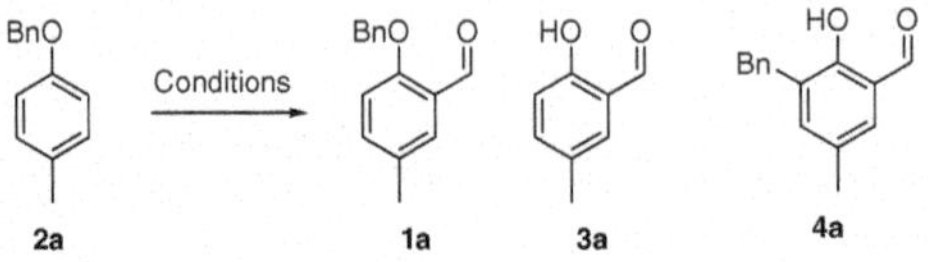

Entry	Reagents (equiv)	Solv.	Temp (°C)	Time	Ratio of **1a**:**3a**[a]	Product (yield, %)[b]
1	$POCl_3$ (5)	NMF	80	12 h	–	N. R.
2	$POCl_3$ (5)	NMFA	120	24 h	–	N. R.
3	HMTA (1.1)	TFA	reflux	1 h	–	**4a** (28)
4	Cl_2CHOMe (3) $TiCl_4$ (3)	CH_2Cl_2	−78	10 min	–	Complex mixture
5	Cl_2CHOMe (3) $SnCl_4$ (3)	CH_2Cl_2	−78	10 min	4:1	**1a**, **3a**[c]
6	Cl_2CHOMe (3) $AlCl_3$ (3)	CH_2Cl_2	−78	10 min	1.8:1	**1a**, **3a**[c]
7	Cl_2CHOMe (3) AgOTf (3)	CH_2Cl_2	−78	10 min	>95:5	**1a** (76)
8	Cl_2CHOMe (3) AgCl (3)	CH_2Cl_2	Rt	1 h	–	N. R.
9	Cl_2CHOMe (3) AgI (3)	CH_2Cl_2	Rt	1 h	–	N. R.
10	Cl_2CHOMe (3) $AgClO_4$ (3)	CH_2Cl_2	−78	1 h	–	Complex mixture
11	Cl_2CHOMe (3) $AgNTf_2$ (3)	CH_2Cl_2	−78	15 min	–	Complex mixture

[a] The ratio was determined by ^{1}H NMR analysis of the crude product
[b] Isolated yield
[c] Inseparable mixture of **1a** and **3a**

provided the undesired formylated phenol **4a** (Table 4.1, entry 3). It is noteworthy that the newly introduced formyl group behaved as a directing group, and the debenzylation of the product under acidic conditions was followed by the capture of benzyl cation in a Friedel-Crafts fashion to provide **4a** (Scheme 4.1). In the Rieche reaction, which uses Cl_2CHOMe as a formylating reagent in the presence of a Lewis acid [8, 9], the use of $TiCl_4$ resulted in a complex mixture, even at temperatures as low as −78 °C (Table 4.1, entry 4). Side reactions were also observed with milder

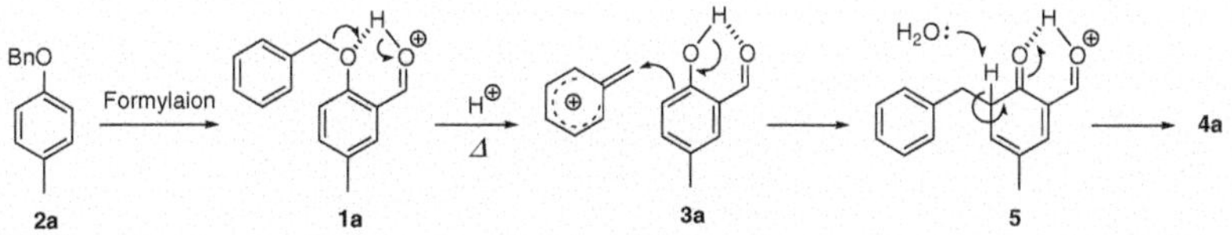

Scheme 4.1 Side reaction of **2a** under the Duff conditions

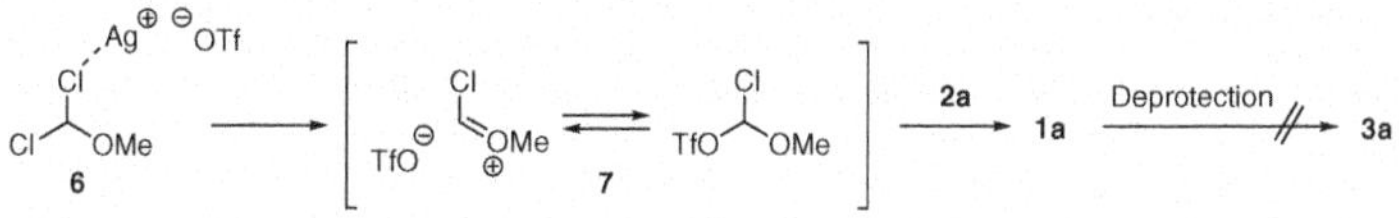

Scheme 4.2 Plausible mechanism for the AgOTf-promoted formylation

Lewis acids such as $SnCl_4$ and $AlCl_3$, which afforded inseparable mixtures of the desired product **1a** and debenzylated material **3a** (Table 4.1, entries 5 and 6). In contrast, the use of AgOTf promoted the formylation of **2a** with Cl_2CHOMe at −78 °C to afford **1a** exclusively without the loss of the benzyl group (Table 4.1, entry 7). In contrast, the use of other Ag salts such as AgCl, AgI, $AgClO_4$ and $AgNTf_2$ was found to be ineffective for the desired formylation (Table 4.1, entries 8–11). These results therefore indicate that AgOTf could be used to specially activate Cl_2CHOMe to furnish the desired formylation product **1a** without the occurrence of any side reactions.

The proposed mechanism for this formylation reaction is shown in Scheme 4.2. The first step in this reaction would most likely involve the activation of Cl_2CHOMe by AgOTf because Ag^+ cations have a high affinity for Cl atoms. Activation of Cl_2CHOMe in this way would lead to the formation of active species **7**, which would be immediately captured by **2a** in a Friedel-Crafts fashion followed by a hydrolysis reaction to provide the corresponding aldehyde **1a**. AgOTf may therefore be considered effective for this formylation in terms of being able to provide highly cationic Ag^+ to activate Cl_2CHOMe and a triflate anion for the stabilization of active species **7**. Furthermore, the resulting TfOH would not be able to affect the removal of the benzyl group in **1a** at such a low temperature.

4.3 Scope and Limitations

With the optimal formylation conditions in hand, we proceeded to evaluate the scope of reaction using a variety of substituted benzenes. Pleasingly, this reaction exhibited good tolerance towards a wide range of phenol-protecting groups, as shown in Table 4.2. Substrates **2b–d**, bearing methyl, allyl and propargyl protected phenols, respectively, reacted smoothly under the optimized conditions to afford the corresponding aldehydes **1b–d** in moderate to good yields (Table 4.2, entries 1–3).

Table 4.2 Formylation of protected 4-methylphenol derivatives

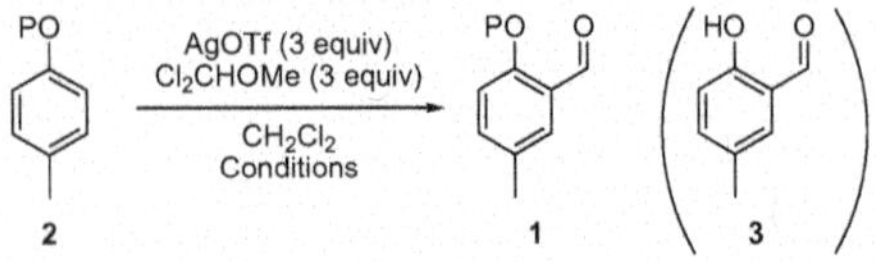

Entry	Substrate	P	Conditions	Product (yield, %)
1	**2b**	Me	−78 °C, 10 min	**1b** (69)
2	**2c**	Allyl	−78 °C, 10 min	**1c** (76)
3	**2d**	Propargyl	−78 °C, 10 min	**1d** (49)
4	**2e**	Ac	0 °C, 12 h	N. R.
5	**2f**	Ts	0 °C, 12 h	N. R.

Notably, compound **1c** was obtained without any side reactions, whereas the desired formylation reaction was accompanied by a Claisen rearrangement when **2c** was treated with Cl_2CHOMe-$TiCl_4$ under conventional Rieche conditions to give phenol **4c** (Scheme 4.3). In contrast, the use of electron-withdrawing protecting groups such as Ac and Ts groups was found to be unsuitable under these conditions, with the formylation reactions of acetate **2e** and tosylate **2f** failing to proceed even at 0 °C (Table 4.2, entries 4 and 5).

To further evaluate the scope and limitation of this new reaction, we proceeded to investigate the formylation of electron-deficient indole derivatives, as shown in Table 4.3. Pleasingly, tosylate **2g** was converted to the corresponding 3-formyl indole derivative **1g** in moderate yield (Table 4.3, entry 1). However, the application of these reaction conditions to indole **2h** bearing an acid-labile Boc group gave a complex mixture of products (Table 4.3, entry 2).

These newly developed formylation conditions were also applied to a variety of different phenol derivatives, as summarized in Table 4.4. Anisole (**2i**) was readily formylated at −78 °C to provide **1i** and its regioisomer **1i′** in a ratio of 1:1.5 (Table 4.4, entry 1). The formylation of 1,3-dimethoxybenzene (**2j**) was also performed at −78 °C, with the least sterically hindered position being formylated to afford **1j** (Table 4.4, entry 2). In contrast, the formylation of 1,3,5-trimethoxybenzene (**2k**) had to be conducted at 0 °C because of steric hindrance, and afforded the monoformylated product **1k** together with a small amount of the diformylated product **8k** (Table 4.4, entry 3). In this particular case, the electron-withdrawing

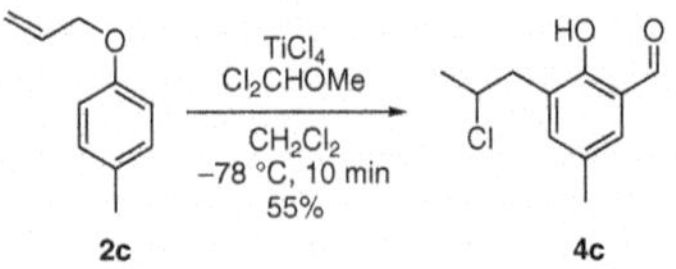

Scheme 4.3 Side reaction of **2c** under the conventional Rieche conditions

Table 4.3 Formylation of protected indole derivatives

AgOTf (3 equiv), Cl_2CHOMe (3 equiv), CH_2Cl_2, Conditions: 2 → 1

Entry	Substrate	P	Conditions	Product (yield, %)
1	**2g**	Ts	−78 °C, 10 min	**1h** (66)
			0 °C, 30 min	
2	**2h**	Boc	−78 °C, 10 min	Complex mixture

Table 4.4 Formylation of phenol derivatives

AgOTf (3 equiv), Cl_2CHOMe (3 equiv), CH_2Cl_2, Conditions: 2 → 1, 1′, 8k

Entry	Substrate	R[1]	R[2]	R[3]	Conditions	Product (yield, %)
1	**2i**	Me	H	H	−78 °C, 10 min	**1i** (28), **1i′** (42)
2	**2j**	Me	OMe	H	−78 °C, 10 min	**1j** (58)
3	**2k**	Me	OMe	OMe	−78 °C, 10 min	**1k** (65), **8k** (4)
4	**2l**	Me	Me	Me	−78 °C, 10 min	**1l** (51), **1l′** (18)
5	**2m**	H	Me	Me	0 °C, 30 min	**1m** (40), **1m′** (15)
6	**2n**	Ac	Me	Me	0 °C, 30 min	**1n** + **1n′** (49)[a]

[a] Inseparable mixture of **1n** and **1n′** was obtained after purification by column chromatography over silica gel. The ratio was determined to be 1.2:1 by ^{1}H NMR analysis

nature of the newly introduced formyl group was not strong enough to overcome the high electron-donating activity of the three methoxy groups, which facilitated a second formylation reaction. Among the 3,5-dimethylphenol derivatives (**2l–n**), methyl ether **2l** was the only substrate successfully formylated at −78 °C to give **1l** and its regioisomer **1l′** in a ratio of 2.8:1 (Table 4.4, entry 4). In contrast, the formylation of phenol **2m** had to be conducted at 0 °C to give **1m** and its regioisomer **1m′**, and the yield was lower than that of **2l** (Table 4.4, entry 5). The partial recovery of **2m** indicated that the *O*-formylation of **2m** may have occurred in competition with *C*-formylation. Acetate **2n**, which was more electron rich than **2e**, underwent the formylation reaction at 0 °C to give an inseparable mixture of **1n** and its regioisomer **1n′** in a ratio of 1.2:1 (Table 4.4, entry 6) [10].

The tolerance of this new method for anisole derivatives bearing electron-withdrawing groups was also investigated, and the results are shown in Table 4.5.

Table 4.5 Formylation of 4-substituted anisole derivatives

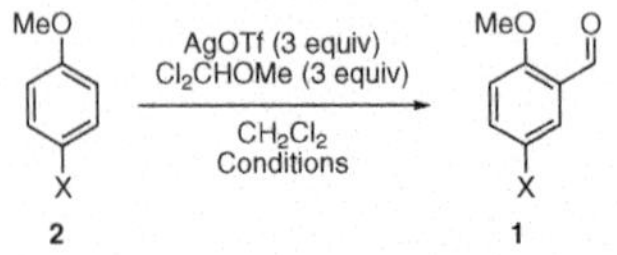

Entry	Substrate	X	Conditions	Product (yield, %)
1	**2o**	Br	−78 °C, 10 min	**1o** (61)
2	**2p**	I	−78 °C, 10 min	**1p** (51)
3	**2q**	COOMe	−78 °C, 10 min	**1q** (66)
			0 °C, 20 min	

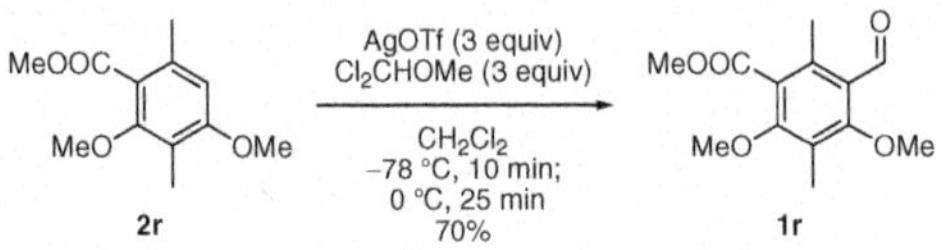

Scheme 4.4 Formylation of pentasubstituted benzene **2r**

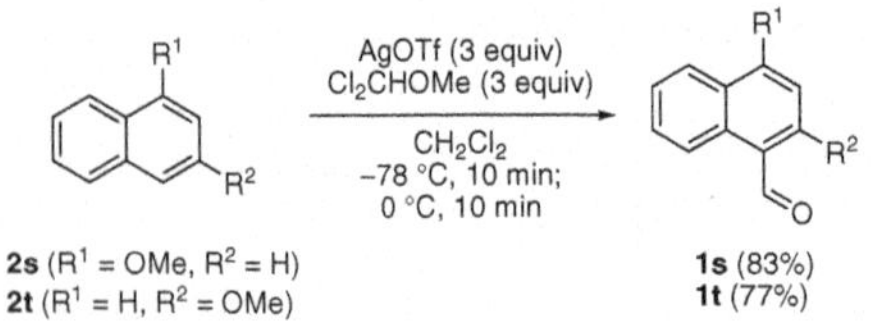

Scheme 4.5 Formylation of methoxynaphthalenes

Although anisoles bearing electron-withdrawing groups such as halogens and methoxycarbonyl groups were found to be less reactive than anisole itself, substrates **2o–q** were formylated to provide the corresponding aldehydes **1o–q** in moderate yields (Table 4.5, entries 1–3).

The formylation of pentasubstituted benzene **2r** with Cl_2CHOMe-AgOTf afforded the fully substituted benzaldehyde **1r** (Scheme 4.4). This result is particularly important from the perspective of the step-economy of the reaction because the corresponding Duff reaction invariably requires free phenols as its substrates [11]. Methoxynaphthalenes **2s** and **2t** were kinetically formylated to afford the corresponding aldehydes **1s** and **1t**, respectively (Scheme 4.5).

Finally, the formylation of alkylbenzenes was investigated using the Cl_2CHOMe-AgOTf conditions (Table 4.6). The electron-rich 1,3,5-trialkylbenzene substrates **2u** and **2v** were readily converted to corresponding aldehydes **1u** and **1v**, respectively (Table 4.6, entries 1 and 2). In contrast, the formylation of substrate **2w**

Table 4.6 Formylation of alkylbenzenes

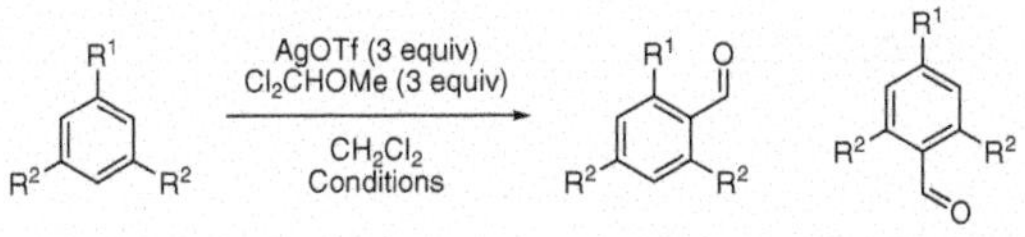

Entry	Substrate	R^1	R^2	Conditions	Product (yield, %)
1	**2u**	Me	Me	−78 °C, 20 min	**1u** (69)
2	**2v**	*i*-Pr	*i*-Pr	−78 °C, 20 min	**1v** (72)
3	**2w**	*t*-Bu	*t*-Bu	0 °C, 12 h	N. R.
4	**2x**	Br	Me	0 °C, 10 min	**1x** (62), **1x′** (18)
5	**2y**	Me	Br	Rt, 1.5 h	**1y** (3), **1y′** (2)

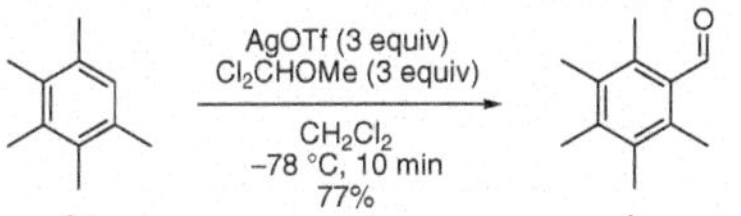

Scheme 4.6 Formylation of pentamethylbenzene (**2z**)

bearing *t*-Bu groups at 1-,3- and 5-positions did not proceed because of the steric hindrance of the *t*-Bu groups (Table 4.6, entry 3). Although monobromobenzene **2x** underwent the formylation at 0 °C to give a 3.3:1 mixture of **1x** and its regioisomer **1x′** (Table 4.6, entry 4), the formylation of dibromobenzene **2y** hardly proceeded, even at room temperature (Table 4.6, entry 5). Notably, the van der Waals radius of a Br atom is almost identical to that of a methyl group, which indicated that these results were caused by electronic factors. Pentamethylbenzene (**2z**), which is known to be a radical scavenger, was also smoothly formylated to afford **1z** (Scheme 4.6).

4.4 Summary

In conclusion, the work described in this chapter demonstrates that Cl_2CHOMe-AgOTf can be used for the formylation of a wide range of substituted benzenes. Substrates were readily formylated at low temperatures to provide the corresponding aldehydes in moderate yields without the loss of their phenol-protecting groups, benzyl, methyl, allyl and propargyl ethers. This atom- and step-economical formylation could therefore be useful for the synthesis of highly functionalized aromatic compounds.

4.5 Experimental Section

4.5.1 General Techniques

All commercially available reagents were used as received. Dry CH_2Cl_2 (Kanto Chemical Co.) were obtained by passing commercially available pre-dried, oxygen-free formulations.

All reactions were monitored by thin-layer chromatography carried out on 0.2 mm E. Merck silica gel plates (60F-254) with UV light, and visualized by p-anisaldehyde H_2SO_4-ethanol solution or phosphomolybdic acid ethanol solution. Flash column chromatography was carried out with silica gel 60 N (Kanto Chemical Co. 40–50 μm).

^{1}H NMR spectra (400 MHz) and ^{13}C NMR spectra (100 MHz) were recorded on JEOL JNM-AL400, in the indicated solvent. Chemical shifts (δ) are reported in units parts per million (ppm) relative to the signal for internal tetramethylsilane (0.00 ppm for ^{1}H) for solutions in chloroform-*d*. NMR spectral data are reported as follows: chloroform-*d* (77.0 ppm for ^{13}C), dimethyl sulfoxide-d_4 (2.49 ppm for ^{1}H and 39.5 ppm for ^{13}C) when internal standard is not indicated. Multiplicities are reported by the following abbreviations: s (singlet), d (doublet), t (triplet), quin (quintet), m (multiplet), dd (double doublet), dt (double triplet), dq (double quartet), ddd (double double doublet), ddt (double double triplet).

EI mass spectra and high-resolution EI mass spectra were measured on JEOL JMS-DX303. IR spectra were recorded on a JASCO FTIR-8400. Only the strongest and/or structurally important absorption are reported as the IR data afforded in cm^{-1}. Melting points were measured on Round Science RFS-10, and are uncorrected.

4.5.2 Experimental Methods

To a solution of *p*-cresol (541 mg, 5.00 mmol) in DMF (10 mL) were added K_2CO_3 (2.07 g, 15.0 mmol, 3.0 equiv) and alkyl bromide (7.50 mmol, 1.5 equiv) at room temperature under an argon atmosphere. After being stirred at the same temperature, the reaction mixture was filtered through a pad of Celite®. The filtrate was diluted with EtOAc, and acidified with 3 M aq HCl. The organic layer was separated, and the aqueous layer was extracted with EtOAc twice. The combined organic layers were washed with brine twice, saturated aq $NaHCO_3$ and brine, dried with $MgSO_4$, and filtered. The filtrate was concentrated in vacuo, and the resulting residue was purified by flash column chromatography on silica gel (eluted with hexane/EtOAc = 20:1) to afford alkoxy benzenes **2a**, **2c** and **2d**.

1-Benzyloxy-4-methylbenzene (2a) [12].
Conditions: BnBr (891 μL), 5.5 h. Yield: benzyl ether **2a** (906 mg, 4.57 mmol, 91 %) as a white solid. mp 37–38 °C [lit. 41–42 °C] [13]; ^{1}H NMR (400 MHz, $CDCl_3$) δ 7.31–7.44 (5H, m), 7.08 (2H, d, J = 8.4 Hz), 6.88 (2H, d, J = 8.4 Hz),

5.04 (2H, s), 2.29 (3H, s); ^{13}C NMR (100 MHz, $CDCl_3$) δ 156.7, 137.3, 130.1, 129.9, 128.5, 127.8, 127.4, 115.8, 114.7, 70.0, 20.5; IR (Neat) 3,031, 2,922, 1,615, 1,585, 1,511, 1,455, 1,239, 1,026, 734, 696 cm^{-1}; HREIMS calcd for $C_{14}H_{14}O$ 198.1045, found 198.1036.

1-Allyloxy-4-methylbenzene (2c) [14].
Conditions: allyl bromide (648 μL), 5.5 h. Yield: allyl ether **2c** (622 mg, 4.20 mmol, 84 %) as a colorless oil. ^{1}H NMR (400 MHz, $CDCl_3$) δ 7.07 (2H, d, J = 8.8 Hz), 6.82 (2H, s, J = 8.8 Hz), 6.01 (1H, ddt, J = 17.2, 10.2, 5.2 Hz), 5.40 (1H, dq, J = 17.2, 1.4 Hz), 5.27 (1H, dq, J = 10.2, 1.4 Hz), 4.51 (2H, dt, J = 5.2, 1.4 Hz), 2.28 (3H, s); ^{13}C NMR (100 MHz, $CDCl_3$) δ 156.4, 133.5, 129.84, 129.77, 117.2, 114.5, 68.7, 20.3; IR (Neat) 3,029, 2,922, 1,613, 1,585, 1,510, 1,291, 1,241, 1,029, 818 cm^{-1}; HREIMS calcd for $C_{10}H_{12}O$ 148.0888, found 148.0887.

1-Methyl-4-(prop-2-yn-1-yloxy)benzene (2d) [15].
Conditions: propargyl bromide (561 μL), 3.5 h. Yield: propargyl ether **2d** (376 mg, 2.57 mmol, 51 %) as a colorless oil. ^{1}H NMR (400 MHz, $CDCl_3$) δ 7.10 (2H, d, J = 8.6 Hz), 6.88 (2H, d, J = 8.6 Hz), 4.66 (2H, d, J = 2.4 Hz), 2.50 (1H, t, J = 2.4 Hz), 2.29 (3H, s); ^{13}C NMR (100 MHz, $CDCl_3$) δ 155.4, 130.8, 129.9, 114.7, 78.8, 75.3, 55.8, 20.4; IR (Neat) 3,291, 2,923, 2,117, 1,609, 1,587, 1,511, 1,218, 1,032, 808 cm^{-1}; HREIMS calcd for $C_{10}H_{10}O$ 146.0732, found 146.0718.

General procedure: Synthesis of protected phenols 2e and 2f.
To a solution of *p*-cresol (300 mg, 2.77 mmol) in dry CH_2Cl_2 (5.0 mL) were added Et_3N (965 μL, 6.93 mmol, 2.5 equiv), protecting reagent (3.33 mmol, 1.2 equiv) and DMAP (16.9 mg, 0.139 mmol, 0.05 equiv) at room temperature under an argon atmosphere. After being stirred at the same temperature for 12 h, the reaction mixture was quenched with 3 M aq HCl. The organic layer was separated, and the aqueous layer was extracted with EtOAc twice. The combined organic layers were washed with saturated aq $NaHCO_3$ and brine, dried with $MgSO_4$, and filtered. The filtrate was concentrated in vacuo, and the resulting residue was purified by flash column chromatography on silica gel (eluted with hexane/EtOAc = 9:1) to afford protected phenols **2e** and **2f**.

4-Methylphenyl acetate (2e) [16].
Conditions: Ac_2O (315 μL). Yield: acetate **2e** (382 mg, 2.54 mmol, 92 %) as a colorless oil. ^{1}H NMR (400 MHz, $CDCl_3$) δ 7.17 (2H, d, J = 8.4 Hz), 6.96 (2H, d, J = 8.4 Hz), 2.34 (3H, s), 2.28 (3H, s); ^{13}C NMR (100 MHz, $CDCl_3$) δ 169.5, 148.5, 135.3, 129.8, 121.2, 20.9, 20.7; IR (Neat) 3,035, 2,925, 1,760, 1,506, 1,369, 1,217, 1,197, 1,166, 909 cm^{-1}; HREIMS calcd for $C_9H_{10}O_2$ 150.0681, found 150.0684.

4-Methylphenyl tosylate (2f) [17].
Conditions: TsCl (635 mg). Yield: tosylate **2f** (629 mg, 2.40 mmol, 87 %) as a white solid. mp 68–69 °C [lit. 68–69 °C] [18]; ^{1}H NMR (400 MHz, $CDCl_3$) δ 7.70 (2H, d, J = 8.6 Hz), 7.30 (2H, d, J = 8.6 Hz), 7.06 (2H, d, J = 8.6 Hz), 6.85 (2H, d, J = 8.6 Hz), 2.44 (3H, s), 2.30 (3H, s); ^{13}C NMR (100 MHz, $CDCl_3$) δ 147.5,

145.2, 136.7, 132.5, 130.0, 129.7, 128.5, 122.0, 21.7, 20.8; IR (Neat) 3,041, 1,597, 1,376, 1,198, 1,175, 1,158, 829, 654 cm^{-1}; HREIMS calcd for $C_{14}H_{14}O_3S$ 262.0664, found 262.0664.

3,5-Dimethylphenyl acetate (2n) [19].
To a solution of 3,5-dimethylphenol (1.00 g, 8.19 mmol) in dry CH_2Cl_2 (10 mL) were added Et_3N (3.41 mL, 24.6 mmol, 3.0 equiv), Ac_2O (1.08 mL, 11.5 mmol, 1.4 equiv) and DMAP (20.0 mg, 0.164 mmol, 0.02 equiv) at room temperature under an argon atmosphere. After being stirred at the same temperature for 2 h, the reaction mixture was quenched with 3 M aq HCl. The organic layer was separated, and the aqueous layer was extracted twice with EtOAc. The combined organic layers were washed with saturated aq $NaHCO_3$ and brine, dried with $MgSO_4$, and filtered. The filtrate was concentrated in vacuo, and the resulting residue was purified by flash column chromatography on silica gel (eluted with hexane/EtOAc = 9:1) to afford acetate **2n** (1.30 g, 7.92 mmol, 97 %) as a colorless oil. ^1H NMR (400 MHz, $CDCl_3$) δ 6.86 (1H, s), 6.70 (2H, s), 2.31 (6H, s), 2.28 (3H, s); ^{13}C NMR (100 MHz, $CDCl_3$) δ 169.4, 150.4, 139.0, 127.3, 119.0, 21.0, 20.8; IR (Neat) 2,921, 1,761, 1,618, 1,591, 1,369, 1,210, 1,137, 1,032, 677 cm^{-1}; HREIMS calcd for $C_{10}H_{12}O_2$ 164.0837, found 164.0823.

Methyl 2,4-dimethoxy-3,6-dimethylbenzoate (2r).
To a solution of methyl 2,4-dihydroxy-3,6-dimethylbenzoate (1.00 g, 5.10 mmol) in DMF (10 mL) were added K_2CO_3 (5.64 g, 40.8 mmol, 8.0 equiv) and MeI (925 μL, 20.4 mmol, 4.0 equiv) at room temperature under an argon atmosphere. After being stirred at 50 °C for 9 h, the reaction mixture was filtered through a pad of Celite®. The filtrate was diluted with EtOAc, and acidified with 3 M aq HCl. The organic layer was separated, and the aqueous layer was extracted with EtOAc twice. The combined organic layers were washed with brine twice, saturated aq $NaHCO_3$ and brine, dried with $MgSO_4$, and filtered. The filtrate was concentrated in vacuo, and the resulting residue was purified by flash column chromatography on silica gel (eluted with hexane/EtOAc = 9:1) to afford methyl ether **2r** (1.12 g, 5.01 mmol, 98 %) as a colorless oil. ^1H NMR (400 MHz, $CDCl_3$) δ 6.46 (1H, s), 3.90 (3H, s), 3.82 (3H, s), 3.75 (3H, s), 2.30 (3H, s), 2.10 (s, 3H, d); ^{13}C NMR (100 MHz, $CDCl_3$) δ 168.8, 159.2, 156.5, 134.5, 120.6, 117.0, 107.6, 61.6, 55.4, 51.8, 19.6, 8.5; IR (Neat) 2,949, 1,733, 1,605, 1,579, 1,464, 1,322, 1,277, 1,154, 1,121 cm^{-1}; HREIMS calcd for $C_{12}H_{16}O_4$ 224.1049, found 224.1052.

3-Benzyl-2-hydroxy-5-methylbenzaldehyde (4a).
To a solution of benzyl ether **2a** (198 mg, 1.0 mmol) in TFA (5 mL) was added HMTA (154 mg, 1.10 mmol, 1.1 equiv) at room temperature under an argon atmosphere. After being stirred at 80 °C for 1 h, the reaction mixture was cooled to room temperature, and concentrated in vacuo. To the resulting residue was poured water (5 mL) at room temperature. After being stirred at 80 °C for 12 h, the reaction mixture was cooled to room temperature, diluted with EtOAc. The organic layer was separated, and the aqueous layer was extracted with EtOAc twice. The combined organic layers were basified with 2 M aq NaOH. The organic layer was

separated, washed with brine, dried with $MgSO_4$, and filtered. The filtrate was concentrated in vacuo, and the resulting residue was purified by flash column chromatography on silica gel (eluted with hexane/EtOAc = 50:1) to afford aldehyde **4a** (62.8 mg, 27.7 μmol, 28 %) as a yellowish solid. mp 74–75 °C; ^1H NMR (400 MHz, $CDCl_3$) δ 11.1 (1H, s), 9.84 (1H, s), 7.16–7.34 (7H, m), 3.99 (2H, s), 2.28 (3H, s); ^{13}C NMR (100 MHz, $CDCl_3$) δ 196.6, 157.4, 140.1, 138.6, 131.6, 129.6, 128.9, 128.8, 128.4, 126.1, 120.1, 34.7, 20.3; IR (Neat) 3,026, 2,923, 2,851, 1,651, 1,452, 1,260 cm^{-1}; HREIMS calcd for $C_{15}H_{14}O_2$ 226.0994, found 226.1005.

3-(2-Chloropropyl)-2-hydroxy-5-methylbenzaldehyde (4c).
To a solution of allyl ether **2c** (1.00 mmol) in dry CH_2Cl_2 (1.5 mL) were added $TiCl_4$ (1.0 mM in CH_2Cl_2, 3.0 mL, 3.00 mmol, 3.0 equiv) and a solution of Cl_2CHOMe (265 μL, 3.00 mmol, 3.0 equiv) in dry CH_2Cl_2 (0.5 mL) at −78 °C under an argon atmosphere. After being stirred at the same temperature for 10 min, the reaction mixture was quenched with 1 M aq HCl. The organic layer was separated, and the aqueous layer was extracted with EtOAc twice. The combined organic layers were washed with saturated aq $NaHCO_3$ and brine, dried with $MgSO_4$, and filtered. The filtrate was concentrated in vacuo, and the resulting residue was purified by flash column chromatography on silica gel (eluted with hexane/EtOAc = 19:1) to afford the aldehyde **4c** (118 mg, 55.2 μmol, 55 %) as a yellowish oil. ^1H NMR (400 MHz, CD_3OD) δ 9.86 (1H, s), 7.40 (1H, d, J = 1.6 Hz), 7.32 (1H, d, J = 1.6 Hz), 4.38 (1H, double quin, J = 7.6, 6.4 Hz), 3.06 (1H, dd, J = 14.2, 6.4 Hz), 2.99 (1H, dd, J = 14.2, 7.6 Hz), 2.32 (3H, s), 1.49 (3H, d, J = 6.4 Hz); ^{13}C NMR (100 MHz, $CDCl_3$) δ 196.7, 157.7, 139.8, 132.3, 128.6, 1,226.4, 120.0, 57.0, 40.3, 25.1, 20.2; IR (Neat) 3,406, 2,977, 2,926, 2,864, 1,652, 1,464, 1,263, 716 cm^{-1}; HREIMS calcd for $C_{11}H_{13}ClO_2$ 212.0604, found 212.0600.

General procedure: Formylation of aromatic compounds 2 utilizing Cl_2CHOMe-AgOTf.
To a suspension of substrates **2** (1.00 mmol) and AgOTf (771 mg, 3.00 mmol, 3.0 equiv) in dry CH_2Cl_2 (1.5 mL) was added a solution of Cl_2CHOMe (265 μL, 3.00 mmol, 3.0 equiv) in dry CH_2Cl_2 (0.5 mL) at −78 °C under an argon atmosphere. After being stirred at the optimal temperature, the reaction mixture was quenched with saturated aq $NaHCO_3$. After being stirred at room temperature for 30 min, the reaction mixture was filtered through a pad of Celite®. The organic layer was separated, and the aqueous layer was extracted with EtOAc twice. The combined organic layers were washed with brine, dried with $MgSO_4$, and filtered. The filtrate was concentrated in vacuo, and the resulting residue was purified by flash column chromatography on silica gel (eluted with hexane/EtOAc) to afford the corresponding aldehydes **1**.

2-Benzyloxy-5-methylbenzaldehyde (1a) [20].
Conditions: **2a** (198 mg), −78 °C, 10 min. Purification: flash column chromatography on silica gel (eluted with hexane/EtOAc = 50:1). Yield: **1a** (172 mg, 760 μmol, 76 %) as a white solid. mp 56–57 °C [lit. 58.5–59 °C] [21]; Rf 0.52

(hexane/EtOAc = 4:1), ^{1}H NMR (400 MHz, $CDCl_3$) δ 10.5 (1H, s), 7.65 (1H, d, J = 2.0 Hz), 7.31–7.44 (6H, m), 6.94 (1H, d, J = 8.4 Hz), 5.15 (2H, s), 2.30 (3H, s); ^{13}C NMR (100 MHz, $CDCl_3$) δ 189.8, 159.1, 136.5, 136.2, 130.4, 128.6, 128.4, 128.2, 127.2, 124.8, 113.0, 70.5, 20.2; IR (Neat) 3,033, 2,923, 2,861, 1,685, 1,612, 1,583, 1,500, 1,286, 1,246, 1,220, 1,160, 1,025, 725, 696 cm^{-1}; HREIMS calcd for $C_{15}H_{14}O_2$ 226.0994, found 226.0978.

2-Methoxy-5-methylbenzaldehyde (1b) [22].
Conditions: 1-methoxy-4-methylbenzene (126 μL), −78 °C, 10 min. Purification: flash column chromatography on silica gel (eluted with hexane/EtOAc = 19:1). Yield: **1b** (89.4 mg, 595 μmol, 60 %) as a yellowish oil. Rf 0.53 (hexane/EtOAc = 4:1); ^{1}H NMR (400 MHz, $CDCl_3$) δ 10.4 (1H, s), 7.63 (1H, d, J = 2.4 Hz), 7.36 (1H, dd, J = 8.4, 2.4 Hz), 6.89 (1H, d, J = 8.4 Hz), 3.91 (3H, s), 2.32 (3H, s); ^{13}C NMR (100 MHz, $CDCl_3$) δ 189.9, 159.9, 136.5, 129.9, 128.4, 124.4, 111.5, 55.6, 20.1; IR (Neat) 2,946, 2,863, 1,680, 1,611, 1,583, 1,497, 1,394, 1,285, 1,254, 1,157, 1,029 cm^{-1}; HREIMS calcd for $C_9H_{10}O_2$ 150.0681, found 150.0670.

2-Allyloxy-5-methylbenzaldehyde (1c) [23].
Conditions: **2c** (148 mg), −78 °C, 10 min. Purification: flash column chromatography on silica gel (eluted with hexane/EtOAc = 50:1). Yield: **1c** (122 mg, 691 μmol, 69 %) as a yellowish oil. Rf 0.53 (hexane/EtOAc = 4:1); ^{1}H NMR (400 MHz, $CDCl_3$) δ 10.5 (1H, s), 7.64 (1H, d, J = 2.4 Hz), 7.33 (1H, dd, J = 8.2, 2.4 Hz), 6.88 (1H, d, J = 8.2 Hz), 6.07 (1H, ddt, J = 17.2, 10.6, 5.0 Hz, c), 5.44 (1H, dq, J = 17.2, 1.4 Hz, a), 5.33 (1H, dq, J = 10.6, 1.4 Hz), 4.63 (1H, dt, J = 5.0, 1.4 Hz), 2.31 (3H, s); ^{13}C NMR (100 MHz, $CDCl_3$) δ 189.8, 159.0, 136.4, 132.5, 130.2, 128.3, 124.7, 117.8, 112.8, 69.2, 20.2; IR (Neat) 2,860, 1,685, 1,612, 1,496, 1,284, 1,247, 1,224, 1,161, 995 cm^{-1}; HREIMS calcd for $C_{11}H_{12}O_2$ 176.0837, found 176.0823.

5-Methyl-2-(prop-2-yn-1-yloxy)-benzaldehyde (1d) [24].
Conditions: **2d** (146 mg), −78 °C, 10 min. Purification: flash column chromatography on silica gel (eluted with hexane/EtOAc = 50:1). Yield: **1d** (85.7 mg, 492 μmol, 49 %) as a white solid. mp 67–68 °C [lit. 66–68 °C]; Rf 0.41 (hexane/EtOAc = 4:1); ^{1}H NMR (400 MHz, $CDCl_3$) δ 10.5 (1H, s), 7.67 (1H, d, J = 2.0 Hz), 7.37 (1H, dd, J = 8.4, 2.0 Hz), 7.02 (1H, d, J = 8.4 Hz), 4.81 (2H, d, J = 2.6 Hz), 2.55 (1H, t, J = 2.6 Hz), 2.33 (3H, s); ^{13}C NMR (100 MHz, $CDCl_3$) δ 189.7, 157.8, 136.3, 131.2, 128.6, 125.3, 113.3, 77.8, 76.3, 56.5, 20.3; IR (Neat) 3,239, 3,251, 2,925, 2,871, 2,123, 1,685, 1,608, 1,583, 1,493, 1,284, 809 cm^{-1}; HREIMS calcd for $C_{11}H_{10}O_2$ 174.0681, found 174.0690.

1-[(4-Methylphenyl)-sulfonyl]-1H-indole-3-carbaldehyde (2g) [25].
Conditions: 1-[(4-methylphenyl)sulfonyl]-1H-indole (271 mg), −78 °C, 10 min, then 0 °C, 30 min. Purification: flash column chromatography on silica gel (eluted with hexane/EtOAc = 3:1). Yield: **2g** (198 mg, 661 μmol, 66 %) as a yellowish solid. mp 146–147 °C [lit. 147–149 °C]; Rf 0.15 (hexane/EtOAc = 4:1); ^{1}H NMR (400 MHz, $CDCl_3$) δ 10.1 (1H, s), 8.26 (1H, d, J = 7.8 Hz), 8.23 (1H, s), 7.95 (1H, dd, J = 7.8, 1.1 Hz), 7.86 (2H, d, J = 8.4 Hz), 7.41 (1H, dt, J = 7.8, 1.3 Hz), 7.36

(1H, dt, J = 7.8, 1.1 Hz), 7.30 (2H, d, J = 8.4 Hz), 2.38 (3H, s); ^{13}C NMR (100 MHz, $CDCl_3$) δ 185.3, 146.1, 136.2, 135.2, 134.4, 130.3, 127.2, 126.3, 125.0, 122.6, 122.4, 113.2, 21.6; IR (Neat) 3,127, 2,824, 1,679, 1,596, 1,541, 1,379, 1,177, 1,100, 970, 748, 661 cm^{-1}; HREIMS calcd for $C_{16}H_{12}NO_3S$ 299.0616, found 299.0615.

2-Methoxybenzaldehyde (1i) [26] and 4-Methoxybenzaldehyde (1i′) [27].
Conditions: anisole (108 μL), −78 °C, 5 min. Purification: flash column chromatography on silica gel (eluted with hexane/EtOAc = 40:1). Yields: **1i** (38.6 mg, 28.4 μmol, 28 %) as an orange oil and **1i′** (59.1 mg, 43.4 μmol, 43 %) as an orange oil. **1i**: Rf 0.27 (hexane/EtOAc = 9:1); ^{1}H NMR (400 MHz, $CDCl_3$) δ 10.5 (1H, s), 7.83 (1H, dd, J = 7.6, 1.6 Hz), 7.56 (1H, ddd, J = 8.4, 7.6, 1.6 Hz), 7.03 (1H, t, J = 7.6 Hz), 6.89 (1H, d, J = 8.4 Hz), 3.93 (3H, s); ^{13}C NMR (100 MHz, $CDCl_3$) δ 189.8, 161.8, 135.9, 128.5, 124.8, 120.6, 111.6, 55.6; IR (Neat) 2,945, 2,845, 1,688, 1,600, 1,484, 1,287, 1,246, 758 cm^{-1}; HREIMS calcd for $C_8H_8O_2$ 136.0524, found 136.0518. **1i′**: Rf 0.19 (hexane/EtOAc = 9:1); ^{1}H NMR (400 MHz, $CDCl_3$) δ 9.89 (1H, s), 7.85 (2H, d, J = 8.8 Hz), 7.01 (2H, d, J = 8.8 Hz), 3.90 (3H, s); ^{13}C NMR (100 MHz, $CDCl_3$) δ 190.8, 164.5, 131.9, 129.9, 114.2, 55.5; IR (Neat) 2,841, 1,684, 1,600, 1,577, 1,511, 1,260, 1,160, 834 cm^{-1}; HREIMS calcd for $C_8H_8O_2$ 136.0524, found 136.0518.

2,4-Dimethoxybenzaldehyde (1j) [28].
Conditions: 1,3-dimethoxybenzene (129 μL), −78 °C, 15 min. Purification: flash column chromatography on silica gel (eluted with hexane/EtOAc = 9:1). Yield: **1j** (96.4 mg, 580 μmol, 58 %) as a white solid. mp 66–67 °C [lit. 69–71 °C] [29]; Rf 0.09 (hexane/EtOAc = 9:1); ^{1}H NMR (400 MHz, $CDCl_3$) δ 10.3 (1H, s), 7.82 (1H, d, J = 8.4 Hz), 6.56 (1H, d, J = 8.4 Hz), 6.45 (1H, s), 3.91 (3H, s), 3.88 (3H, s); ^{13}C NMR (100 MHz, $CDCl_3$) δ 188.3, 166.2, 163.6, 130.8, 119.1, 105.7, 97.9, 55.61, 55.59; IR (Neat) 2,977, 2,863, 2,781, 1,673, 1,600, 1,580, 1,456, 1,335, 1,285, 1,268, 1,216, 1,028, 829 cm^{-1}; HREIMS calcd for $C_9H_{10}O_3$ 166.0630, found 166.0616.

2,4,6-Trimethoxybenzaldehyde (1k) [30] and 2,4-Diformyl-1,3,5-trimethoxybenzene (8k) [31].
Conditions: 1,3,5-trimethoxybenzene (168 mg), 0 °C, 15 min. Purification: flash column chromatography on silica gel (eluted with hexane/EtOAc = 1:2 to 1:4). Yields: **1k** (127 mg, 647 μmol, 65 %) as a white solid and **8k** (7.9 mg, 35.2 μmol, 4 %) as a white solid. **1k**: mp 132–133 °C, [lit. 119–121 °C]; Rf 0.18 (hexane/EtOAc = 9:1); ^{1}H NMR (400 MHz, $CDCl_3$) δ 10.4 (1H, s), 6.08 (2H, s), 3.89 (6H, s), 3.88 (3H, s); ^{13}C NMR (100 MHz, $CDCl_3$) δ 188.7, 166.2, 164.1, 108.8, 90.2, 56.0, 55.5; IR (Neat) 2,975, 2,881, 2,843, 2,796, 1,671, 1,606, 1,578, 1,475, 1,334, 1,230, 1,217, 1,129, 809 cm^{-1}; HREIMS calcd for $C_{10}H_{12}O_4$ 196.0736, found 196.0729. **8k**: mp 169–170 °C; Rf 0.30 (hexane/EtOAc = 1:4); ^{1}H NMR (400 MHz, $CDCl_3$) δ 10.3 (2H, s), 6.28 (1H, s), 4.01 (6H, s), 3.96 (3H, s); ^{13}C NMR (100 MHz, $CDCl_3$) δ 187.1, 167.9, 167.2, 112.7, 90.9, 64.9, 56.3; IR (Neat)

2,953, 2,859, 1,679, 1,589, 1,559, 1,236, 1,149, 1,106 cm^{-1}; HREIMS calcd for $C_{11}H_{12}O_5$ 224.0685, found 224.0674.

4,6-Dimethyl-2-methoxybenzaldehyde (1l) [32] and 2,6-Dimethyl-4-methoxybenzaldehyde (1l′) [33].
Conditions: 1,3-dimethoxy-5-methylbenzene (142 μL), −78 °C, 10 min. Purification: flash column chromatography on silica gel (eluted with hexane/EtOAc = 50:1). Yields: **1l** (84.0 mg, 512 μmol, 51 %) as a white solid and **1l′** (28.8 mg, 175 μmol, 18 %) as a white solid. **1l**: mp 48–49 °C [lit. 48–49 °C]; Rf 0.30 (hexane/EtOAc = 9:1); ^{1}H NMR (400 MHz, $CDCl_3$) δ 10.6 (1H, s), 6.64 (1H, s), 6.63 (1H, s), 3.88 (3H, s), 2.55 (3H, s), 2.35 (3H, s); ^{13}C NMR (100 MHz, $CDCl_3$) δ 191.7, 163.3, 145.6, 142.0, 125.0, 121.0, 109.7, 55.7, 22.1, 21.4; IR (Neat) 2,965, 2,926, 1,678, 1,599, 1,319, 1,148 cm^{-1}; HREIMS calcd for $C_{10}H_{12}O_2$ 164.0837, found 164.0829. **1l′**: mp 42–43 °C [lit. 40–41 °C] [34]; Rf 0.26 (hexane/EtOAc = 9:1); ^{1}H NMR (400 MHz, $CDCl_3$) δ 10.5 (1H, s), 6.59 (2H, s), 3.84 (3H, s), 2.61 (6H, s); ^{13}C NMR (100 MHz, $CDCl_3$) δ 191.6, 162.7, 144.5, 125.9, 114.8, 55.2, 21.0; IR (Neat) 2,961, 2,923, 1,678, 1,609, 1,462, 1,304, 1,202, 1,097, 832 cm^{-1}; HREIMS calcd for $C_{10}H_{12}O_2$ 164.0837, found 164.0824.

2-Hydroxy-4,6-dimethylbenzaldehyde (1m) [35] and 4-Hydroxy-2,6-dimethylbenzaldehyde (1m′) [36].
Conditions: 3,5-Dimethylphenol (122 mg), 0 °C, 30 min, Purification: flash column chromatography on silica gel (eluted with hexane/EtOAc = 30:1 to 4:1). Yields: **1m** (60.1 mg, 402 μmol, 40 %) as a white solid, **1m′** (22.6 mg, 150 μmol, 15 %) as a white solid. **1m**: mp 49–50 °C [lit. 49 °C] [9]; Rf: 0.43 (hexane/EtOAc = 4:1); ^{1}H NMR (400 MHz, $CDCl_3$) δ 12.0 (1H, s), 10.2 (1H, s), 6.63 (1H, s), 6.54 (1H, s), 2.56 (3H, s), 2.31 (3H, s); ^{13}C NMR (100 MHz, $CDCl_3$) δ 194.5, 163.4, 149.2, 141.8, 123.1, 116.5, 116.1, 22.1, 17.9; IR (Neat) 3,412, 2,928, 2,884, 1,641, 1,572, 1,443, 1,311, 1,238, 1,193, 1,038, 804, 757 cm^{-1}; HREIMS calcd for $C_9H_{10}O_2$ 150.0681, found 150.0668. **1m′**: mp 194–195 °C [lit. 190–191 °C] [37]; Rf 0.21 (hexane/EtOAc = 4:1); ^{1}H NMR (400 MHz, DMSO-d_6) δ 10.3 (1H, s), 6.52 (2H, s), 3.34 (6H, s); ^{13}C NMR (100 MHz, DMSO-d_6) δ 191.4, 161.4, 144.1, 124.3, 116.3, 20.5; IR (Neat) 3,132, 2,961, 2,931, 1,652, 1,603, 1,560, 1,315, 1,272, 1,157, 641 cm^{-1}; HREIMS calcd for $C_9H_{10}O_2$ 150.0681, found 150.0689.

2-Acetoxy-4,6-dimethylbenzylaldehyde (1n) and 4-Acetoxy-2,6-dimethyl-benzaldehyde (1n′).
Conditions: **2n** (164 mg), 0 °C, 30 min. Purification: flash column chromatography on silica gel (eluted with hexane/EtOAc = 50:1 to 20:1). Yields: **1n** and **1n′** [10] (95.0 mg, 494 μmol, 49 %, **1n**:**1n′** = 1.2:1) as a colorless oil. **1n**: Rf 0.20 (hexane/EtOAc = 9:1); ^{1}H NMR (400 MHz, $CDCl_3$) δ 10.3 (1H, s), 6.95 (1H, s), 6.80 (1H, s), 2.60 (3H, s), 2.36 (3H, s), 2.35 (3H, s); ^{13}C NMR (100 MHz, $CDCl_3$) δ 189.3, 169.5, 152.7, 145.7, 142.3, 130.4, 123.6, 121.7, 21.6, 20.8, 20.2; IR (Neat) 2,926, 1,773, 1,691, 1,618, 1,369, 1,202, 1,140, 1,050 cm^{-1}; HREIMS calcd for $C_{11}H_{12}O_3$ 192.0786, found 192.0782. **1n′**: Rf 0.20 (hexane/EtOAc = 9:1); ^{1}H NMR (400 MHz, $CDCl_3$) δ 10.6 (1H, s), 6.85 (2H, s), 2.61 (6H, s), 2.31 (3H, s); ^{13}C

NMR (100 MHz, $CDCl_3$) δ 192.2, 168.9, 153.5, 143.5, 130.2, 122.6, 21.1, 20.7; IR (Neat) 2,927, 1,771, 1,683, 1,596, 1,199, 1,134 cm^{-1}; HREIMS calcd for $C_{11}H_{12}O_3$ 192.0786, found 192.0768.

5-Bromo-2-methoxybenzaldehyde (1o) [38].
Conditions: 1-bromo-4-methoxybenzene (125 μL), −78 °C, 10 min. Purification: flash column chromatography on silica gel (eluted with hexane/EtOAc = 9:1). Yield: **1o** (132 mg, 612 μmol, 61 %) as a white solid. mp 116–117 °C [lit. 116–119 °C]; Rf 0.30 (hexane/EtOAc = 4:1); ^1H NMR (400 MHz, $CDCl_3$) δ 10.4 (1H, s), 7.93 (1H, d, J = 2.6 Hz), 7.64 (1H, dd, J = 8.8, 2.6 Hz), 6.90 (1H, d, J = 8.8 Hz, e), 3.93 (3H, s); ^{13}C NMR (100 MHz, $CDCl_3$) δ 188.3, 160.7, 138.3, 131.0, 126.1, 113.7, 113.5, 56.0; IR (Neat) 3,103, 2,967, 2,844, 1,674, 1,590, 1,477, 1,389, 1,266, 1,243, 1,178, 1,019, 823, 756 cm^{-1}; HREIMS calcd for $C_8H_7BrO_2$ 213.9629, found 213.9597.

5-Iodo-2-methoxybenzaldehyde (1p) [39].
Conditions: 1-iodo-4-methoxybenzene (234 mg), −78 °C, 10 min. Purification: flash column chromatography on silica gel (eluted with hexane/EtOAc = 50:1). Yield: **1p** (133 mg, 506 μmol, 51 %) as a white solid. mp 144–145 °C [lit. 142–143 °C]; Rf 0.32 (hexane/EtOAc = 9:1); ^1H NMR (400 MHz, $CDCl_3$) δ 10.3 (1H, s), 8.10 (1H, d, J = 2.4 Hz, c), 7.81 (1H, dd, J = 8.8, 2.4 Hz), 6.79 (1H, d, J = 8.8 Hz), 3.92 (3H, s); ^{13}C NMR (100 MHz, $CDCl_3$) δ 188.2, 161.4, 144.1, 137.0, 126.5, 114.1, 83.0, 55.8; IR (Neat) 2,963, 1,671, 1,584, 1,472, 1,389, 1,268, 1,244, 1,176, 1,020, 819 cm^{-1}; HREIMS calcd for $C_8H_7IO_2$ 261.9491, found 261.9498.

Methyl 3-formyl-4-methoxybenzoate (1q).
Conditions: Methyl 4-methoxybenzoate (166 mg), −78 °C, 10 min, then 0 °C, 20 min. Purification: flash column chromatography on silica gel (eluted with hexane/EtOAc = 4:1). Yield: **1q** (128 mg, 657 μmol, 66 %) as a white solid. mp 101–102 °C; Rf 0.17 (hexane/EtOAc = 4:1); ^1H NMR (400 MHz, $CDCl_3$) δ 10.5 (1H, s), 8.51 (1H, d, J = 2.2 Hz), 8.25 (1H, dd, J = 9.0, 2.2 Hz), 7.05 (1H, d, J = 9.0 Hz), 4.01 (3H, s), 3.91 (3H, s); ^{13}C NMR (100 MHz, $CDCl_3$) δ 188.8, 165.9, 164.7, 137.1, 130.6, 124.4, 122.9, 111.5, 56.0, 52.1; IR (Neat) 2952, 1714, 1,685, 1,606, 1,267, 1,125, 761 cm^{-1}; HREIMS calcd for $C_{10}H_{10}O_4$ 194.0579, found 194.0572.

Methyl 4,6-dimethoxy-2,5-dimethyl-3-formylbenzoate (1r) [40].
Conditions: **2r** (224 mg), −78 °C, 10 min, then 0 °C, 25 min. Purification: flash column chromatography on silica gel (eluted with hexane/EtOAc = 4:1). Yield: **1r** (177 mg, 703 μmol, 70 %) as a yellowish oil. Rf 0.26 (hexane/EtOAc = 4:1); ^1H NMR (400 MHz, $CDCl_3$) δ 10.4 (1H, s), 3.94 (3H, s), 3.84 (3H, s), 3.82 (3H, s), 2.47 (3H, s), 2.23 (3H, s); ^{13}C NMR (100 MHz, $CDCl_3$) δ 191.5, 168.1, 165.1, 160.3, 137.1, 127.2, 124.2, 123.1, 63.1, 61.7, 52.4, 17.2, 8.9; IR (Neat) 2,950, 1,735, 1,685, 1,570, 1,310, 1,206, 1,106 cm^{-1}; HREIMS calcd for $C_{13}H_{16}O_5$ 252.0998, found 252.0991.

4-Methoxynaphthalene-1-carbaldehyde (1s) [33].
Conditions: 1-methoxynaphtalene (158 mg), −78 °C, 10 min, then 0 °C, 10 min. Purification: flash column chromatography on silica gel (eluted with hexane/EtOAc = 9:1). Yield: **1s** (154 mg, 828 μmol, 83 %) as a yellowish oil. Rf: 0.16 (hexane/EtOAc = 9:1); ^{1}H NMR (400 MHz, $CDCl_3$) δ 10.2 (1H, s), 9.31 (1H, d, J = 8.4 Hz), 8.34 (1H, d, J = 8.7 Hz), 7.93 (1H, d, J = 8.0 Hz), 7.70 (1H, ddd, J = 8.4, 7.0, 1.2 Hz), 7.58 (1H, ddd, J = 8.7, 7.0, 1.2 Hz), 6.93 (1H, d, J = 8.0 Hz), 4.11 (3H, s); ^{13}C NMR (100 MHz, $CDCl_3$) δ 192.2, 160.7, 139.6, 131.8, 129.4, 126.3, 125.4, 124.9, 124.8, 122.3, 102.8, 55.9; IR (Neat) 2,940, 2,846, 1,677, 1,619, 1,513, 1,429, 1,251, 1,220, 1,092, 1,059, 765 cm^{-1}; HREIMS calcd for $C_{12}H_{10}O_2$ 186.0681, found 186.0669.

2-Methoxynaphthalene-1-carbaldehyde (1t) [41].
Conditions: 2-methoxynaphtalene (158 mg), −78 °C, 10 min, then 0 °C, 10 min. Purification: flash column chromatography on silica gel (eluted with hexane/EtOAc = 4:1). Yield: **1t** (143 mg, 76.6 μmol, 77 %) as a yellowish solid. mp 83–84 °C [lit. 82.2–83.7 °C]; Rf 0.20 (hexane/EtOAc = 9:1); ^{1}H NMR (400 MHz, $CDCl_3$) δ 10.9 (1H, s), 9.28 (1H, d, J = 8.5 Hz), 8.07 (1H, d, J = 8.8 Hz), 7.78 (1H, d, J = 8.6 Hz), 7.63 (1H, ddd, J = 8.8, 6.9, 1.1 Hz), 7.42 (1H, ddd, J = 8.5, 6.9, 1.7 Hz), 7.31 (1H, d, J = 8.5 Hz), 4.06 (3H, s); ^{13}C NMR (100 MHz, $CDCl_3$) δ 192.0, 163.9, 137.5, 131.5, 129.9, 128.5, 128.2, 124.9, 124.7, 116.6, 112.5, 56.5; IR (Neat) 2,944, 2,888, 1,665, 1,591, 1,513, 1,267, 1,250, 1,149, 804, 749 cm^{-1}; HREIMS calcd for $C_{12}H_{10}O_2$ 186.0681, found 186.0697.

2,4,6-Trimethylbenzaldehyde (1u) [42].
Conditions: mesitylene (138 μL), −78 °C, 20 min. Purification: flash column chromatography on silica gel (eluted with hexane/EtOAc = 19:1). Yield: **1u** (102 mg, 690 μmol, 69 %) as a yellowish oil. Rf 0.35 (hexane/EtOAc = 9:1); ^{1}H NMR (400 MHz, $CDCl_3$) δ 10.6 (1H, s), 6.89 (2H, s), 2.57 (6H, s), 2.31 (3H, s); ^{13}C NMR (100 MHz, $CDCl_3$) δ 192.9, 143.8, 141.4, 130.5, 129.9, 21.4, 20.4; IR (Neat) 2,963, 2,922, 2,863, 1,683, 1,609, 1,436, 1,208, 1,148, 852, 782 cm^{-1}; HREIMS calcd for $C_{10}H_{12}O$ 148.0888, found 148.0873.

2,4,6-Triisopropylbenzaldehyde (1v) [43].
Conditions: 1,3,5-triisopropylbenzene (240 μL), −78 °C, 20 min. Purification: flash column chromatography on silica gel (eluted with hexane/EtOAc = 19:1). Yield: **1v** (168 mg, 723 μmol, 72 %) as a yellowish oil. Rf 0.53 (hexane/EtOAc = 9:1); ^{1}H NMR (400 MHz, $CDCl_3$) δ 10.7 (1H, s), 7.11 (2H, s), 3.60 (2H, septet, J = 6.8 Hz), 2.92 (1H, septet, J = 6.8 Hz), 1.274 (12H, d, J = 6.8 Hz), 1.266 (6H, d, J = 6.8 Hz); ^{13}C NMR (100 MHz, $CDCl_3$) δ 195.0, 153.6, 150.4, 121.6, 34.7, 28.7, 24.2, 23.7; IR (Neat) 2,964, 1,691, 1,604, 1,459, 878 cm^{-1}; HREIMS calcd for $C_{16}H_{24}O$ 232.1827, found 232.1823.

2-Bromo-4,6-dimethylbenzaldehyde (1x) and 4-Bromo-2,6-dimethylbenz-aldehyde (1x′) [44].
Conditions: 1-bromo-3,5-dimethylbenzene (137 μL), 0 °C, 10 min. Purification: flash column chromatography on silica gel (eluted with hexane/EtOAc = 50:1).

Yields: **1x** (132 mg, 621 μmol, 62 %) as a white solid and **1x′** (39.7 mg, 186 μmol, 19 %) as a white solid. **1x**: mp 39–40 °C; Rf 0.40 (hexane/EtOAc = 9:1); ^{1}H NMR (400 MHz, $CDCl_3$) δ 10.5 (1H, s), 7.35 (1H, s), 7.01 (1H, s), 2.56 (3H, s), 2.35 (3H, s); ^{13}C NMR (100 MHz, $CDCl_3$) δ 194.2, 144.9, 142.6, 132.3, 132.2, 129.1, 128.7, 21.24, 21.16; IR (Neat) 2970, 2,927, 2,858, 2,761, 1,691, 1,601, 1,376, 1,131, 848 cm^{-1}; HREIMS calcd for C_9H_9BrO 211.9837, found 211.9822. **1x′**: mp 66–67 °C; Rf 0.36 (hexane/EtOAc = 9:1); ^{1}H NMR (400 MHz, $CDCl_3$) δ 10.6 (1H, s), 7.27 (2H, s), 2.59 (6H, s); ^{13}C NMR (100 MHz, $CDCl_3$) δ 192.5, 143.0, 132.5, 131.1, 127.7, 20.3; IR (Neat) 2,964, 2,925, 1,690, 1,577, 1,417, 1,256, 852 cm^{-1}; HREIMS calcd for C_9H_9BrO 211.9837, found 211.9824.

2,4-Dibromo-6-methylbenzaldehyde (1y) and 2,6-Dibromo-4-methylbenz-aldehyde (1y′) [45].
Conditions: 1,3-dibromo-5-dimethylbenzene (137 μL), room temperature, 1.5 h. Purification: flash column chromatography on silica gel (eluted with hexane/EtOAc = 40:1). Yields: **1y** (7.7 mg, 27.7 μmol, 3 %) as a white solid and **1y′** (5.1 mg, 18.4 μmol, 2 %) as a white solid. **1y**: mp 58–59 °C; Rf 0.41 (hexane/EtOAc = 9:1); ^{1}H NMR (400 MHz, $CDCl_3$) δ 10.5 (1H, s), 7.70 (1H, d, J = 1.6 Hz), 7.39 (1H, d, J = 1.6 Hz), 2.57 (3H, s); ^{13}C NMR (100 MHz, $CDCl_3$) δ 193.6, 144.0, 134.4, 134.1, 130.5, 128.7, 127.7, 21.1; IR (Neat) 2,927, 2,869, 1,699, 1,573, 1,537, 1,379, 1,170, 892, 856, 791 cm^{-1}; HREIMS calcd for $C_8H_6Br_2O$ 275.8785, found 275.8776. **1y′**: mp 100–101 °C [lit. 95–97 °C] [46]; Rf 0.37 (hexane/EtOAc = 9:1); ^{1}H NMR (400 MHz, $CDCl_3$) δ 10.2 (1H, s), 7.48 (2H, s), 2.37 (3H, s); ^{13}C NMR (100 MHz, $CDCl_3$) δ 190.9, 145.6, 134.4, 129.6, 125.1, 20.9; IR (Neat) 2,924, 2,865, 2,761, 1,706, 1,587, 1,058, 858, 733 cm^{-1}; HREIMS calcd for $C_8H_6Br_2O$ 275.8785, found 275.8795.

Pentamethylbenzaldehyde (1z).
Conditions: pentamethylbenzene (148 mg), −78 °C, 10 min. Purification: flash column chromatography on silica gel (eluted with hexane/EtOAc = 9:1). Yield: **1z** (136 mg, 773 μmol, 77 %) as a white solid. mp 150–151 °C [lit. 143–148.5 °C] [47]; Rf 0.41 (hexane/EtOAc = 9:1); ^{1}H NMR (400 MHz, $CDCl_3$) δ 10.6 (1H, s), 2.42 (6H, s), 2.29 (3H, s), 2.24 (6H, s); ^{13}C NMR (100 MHz, $CDCl_3$) δ 196.5, 140.0, 134.5, 133.6, 133.0, 17.6, 16.1; IR (Neat) 2,921, 2,868, 1,688, 1,566, 1,287, 755 cm^{-1}; HREIMS calcd for $C_{12}H_{16}O$ 176.1201, found 176.1181.

References

1. Keith JM. Tetrahedron Lett. 2004;45:2739–42.
2. Fletcher S, Gunning PT. Tetrahedron Lett. 2008;49:4817–9.
3. Deprotection utilizing cheating effects under acidic conditions has been also mentioned in sections 2.3.2, 2.3.3 and 3.4.2.
4. Sörgel S, Tokunaga N, Sasaki K, Okamoto K, Hayashi T. Compound 1a has been synthesized by the formylation via aryl lithium spices. Org Lett. 2008;10:589–92.
5. Vilsmeier A, Haack A. Ber Deutsch Chem Ges. 1927;60:119–22.

6. Downie IM, Earle MJ, Heaney H, Shunhaibar KF. Tetrahedron. 1993;49:4015–34.
7. Smith WE. J Org Chem. 1972;37:3972–3.
8. Rieche A, Gross H, Höft E. Chem Ber. 1960;93:88–94.
9. Gross H, Rieche A, Mattey G. Chem Ber. 1963;96:308–13.
10. After the removal of acetyl group under basic methanolysis, resulting **1m** and **1m′** could be separated by flash column chromatography on silica gel. **1n** and **1n′** were characterized by acetylation of **1m** and **1m′**, respectively.
11. In the total synthesis of thielocin B1, Duff reaction was used after the removal of benzyl group. The detail is shown in section 2.3.1.
12. Mukaiyama T, Shintou T. J Am Chem Soc. 2004;126:7359–67.
13. Curtin DY, Wilhelm M. J Org Chem. 1958;23:9–12.
14. Wang EC, Hsu MK, Lin YL, Huang KS. Heterocycles. 2002;67:1997–2010.
15. Efe C, Lykakis IN, Stratakis M. Chem Commun. 2011;47:803–5.
16. Xi Z, Hao W, Wang P, Cai M. Molecule. 2009;14:3528–37.
17. Sakurai N, Mukaiyama T. Heterocycles. 2007;74:771–90.
18. Lange PP, Linder C. Angew Chem Int Ed. 2010;49:1111–4.
19. Battaini G, Monzani E, Perotti A, Para C, Casella L, Santagostini L, Gullotti M, Dillinger R, Nather C, Tuczek F. J Am Chem Soc. 2003;125:4185–98.
20. Mori K, Kawasaki T, Sueoka S, Akiyama T. Org Lett. 2010;12:1732–5.
21. Ardis AE, Baltzly R, Schoen W. J Am Chem Soc. 1946;58:591–5.
22. Trivedi SV, Mamdapur VR, Indian J. Chem. Sect. B. 1990;29B:876–8.
23. Miege F, Meyer C, Cossy J. Angew Chem Int Ed. 2011;50:5932–7.
24. Sze EML, Rao W, Koh MJ, Chan PWH. Chem Eur J. 2011;17:1437–41.
25. Couladouros EA, Magos AD. Mol. Divers. 2005;9:99–109.
26. Lin C-K, Lu T-J. Tetrahedron. 2010;66:9688–93.
27. Velusamy S, Ahamed M, Punniyamurthy T. Org Lett. 2004;6:4821–4.
28. Rolfe A, Probst DA, Volp KA, Omar I, Flynn DL, Hanson PR. J Org Chem. 2008;73:8785–90.
29. Rivero IA, Espinoza KA, Ochoa A. J Comb Chem. 2004;6:270–4.
30. Gallardo-Goday A, Fierro A, McLean TH, Castillo M, Cassels BK, Reyes-Parada M, Nichols DE. J Med Chem. 2005;48:2407–19.
31. Kuhnert N, Rossignolo GM, Lopez-Periago A. Org Biomol Chem. 2003;1:1157–70.
32. Saba G, Lai A, Monduzzi M. J. Chem. Soc. Perkin Trans II. 1983;1569–72.
33. Tietze L, Vock CA, Krimmelbem IK, Nacke L. Synthesis. 2009;12:2040–60.
34. Kende AS, Koch K, Smith CA. J Am Chem Soc. 1988;110:2210–8.
35. Knight PD, Clarkson G, Hammond ML, Kimberley BS, Scott P. J Organomet Chem. 2005;690:5125–44.
36. Yamada K, Toyota T, Ishimaru M, Sugawara T. New J Chem. 2001; 25:667–9.
37. Davis CT, Geissmas TA. J Am Chem Soc. 1954;76:3507–11.
38. Dabrowski M, Kubicka J, Lulinski S, Serwatowski J. Tetrahedron. 2005;61:6590–5.
39. Yang H, Li Y, Jiang M, Wang J, Fu H. Chem Eur J. 2011;17:5652–60.
40. Gunzinger J, Tabacchi R. Helv Chim Acta. 1985;68:1940–7.
41. Jang K, Miura K, Koyama Y, Tanaka T. Org Lett. 2012;14:3088–91.
42. Fergus S, Eustace SJ, Hegarty AF. J Org Chem. 2004;69:4663–9.
43. Casarini D, Lunazzi L, Mazzanti A. J Org Chem. 2008;73:2811–8.
44. Kumar RJ, Karlsson S, Streich D, Jensen AR, Jäger M, Becker H-C, Bergquist J, Johansson O, Hammarström L. Chem Eur J. 2010;16:2830–42.
45. Luliński S, Serwatowski J. J Org Chem. 2003;68:5337–84.
46. Gunzinger J, Tabacchi R. Helv Chim Acta. 1985;68:1940–7.
47. Smith LI, Nichols J. J Org Chem. 1941;6:489–506.

Chapter 5
Conclusions

Thielocin B1 has been identified as a protein–protein interaction (PPI) inhibitor of proteasome-assembling chaperone 3 (PAC3), and the first total synthesis of thielocin B1 and its spin-labeled derivative have been described in detail. Furthermore, the mechanism of this inhibition process has been thoroughly investigated using a combination of NMR experiments and in silico docking studies.

Chapter 1 highlighted the unique challenges faced by drug discovery scientists targeting PPIs. Natural products generally occupy a broader area of chemical space than the compounds typically found in traditional high-throughput screening libraries, and can potentially be used as tool compounds to identify suitable lead compounds for the development of PPI inhibitors. In this way, thielocin B1 was identified as an inhibitor of PPIs involved in the formation of PAC3 homodimer. Additionally, the mode of interaction between thielocin B1 and PAC3 appears in many ways to be different from that of conventional PPI inhibitors, and further highlights the significance of our work with this natural product.

Chapter 2 provided a detailed account of the first total synthesis of thielocin B1. The 2,2′,6,6′-tetrasubstituted diphenyl ether moiety of thielocin B1 was synthesized from a depsidone skeleton by the chemoselective reduction of its lactone moiety while maintaining its methyl ester. Interestingly, the protecting group on the phenolic hydroxyl group of the depsidone skeleton had a significant influence on the reactivity of this reduction. During the process of elongating the wings to the left and right of this central core, we developed an efficient process for the formylation of the sterically hindered aromatic core using Cl_2CHOMe–AgOTf. NMR chemical shift perturbation experiments involving a ^{15}N-labeled PAC3 homodimer were conducted both in the presence and absence of the synthesized thielocin B1, and significant changes were observed in the chemical shifts of eight of the amino acid residues on the interface of PAC3 homodimer. These data therefore indicated that thielocin B1 promotes the dissociation to monomeric PAC3 through its interaction with PAC3 homodimer.

A detailed description of the synthesis of spin-labeled thielocin B1 was provided in Chap. 3. To obtain detailed information regarding the binding site of thielocin

K. Ohsawa, *Total Synthesis of Thielocin B1 as a Protein–Protein Interaction Inhibitor of PAC3 Homodimer*, Springer Theses, DOI 10.1007/978-4-431-55447-9_5

B1, a spin-labeling reagent was introduced to the wing to the left-hand side of thielocin B1. The effect of paramagnetic relaxation enhancement (PRE) on PAC3 was subsequently observed in the presence of the spin-labeled molecular probe, and distinct decreases were observed in the NMR signal intensities of 16 of the residues of PAC3 homodimer. These differences in the intensities were determined by measuring the ^{1}H–^{15}N HSQC spectra of PAC3 homodimer in the presence of the spin-labeled molecular probe both before and after quenching the radical moiety. An additional in silico docking study was conducted based on the results of these NMR experiments, which suggested that the left wing of thielocin B1 initially interacts with the interface of PAC3 homodimer, and that the right wing of thielocin B1 then moves towards the β-sheet layers of PAC3 to induce the dissociation of PAC3 homodimer.

Chapter 4 provided an overview of our research towards the development of an atom- and step-economical protocol for the formylation of various substituted benzenes using Cl_2CHOMe–AgOTf. This new method allows for the formylation of substrates at low temperatures to give the corresponding aldehydes in moderate yields. Additionally, various phenol-protecting groups, including benzyl, methyl, allyl and propargyl ethers were tolerated by this reaction.

This chapter provided a summary of this research.

Curriculum Vitae

Kosuke Ohsawa

Office Address: Department of Chemical Biology, Max-Planck Institute of Molecular Physiology, Otto-Hahn-Str. 11, 44227, Dortmund, Germany
E-mail: Kosuke.Osawa@mpi-dortmund.mpg.de
Nationality: Japanese

Work Experience

Department of Chemical Biology, Max-Planck Institute of Molecular Physiology
Postdoctoral fellow: May, 2014–present
Supervisor: Prof. Dr. Herbert Waldmann
Graduate School of Pharmaceutical Sciences, Tohoku University
Research associate: April, 2014
Supervisor: Prof. Dr. Takayuki Doi

Education

Graduate School of Pharmaceutical Sciences, Tohoku University
Ph.D: April, 2011–March, 2014
M.S.: April, 2009–March, 2011
Supervisor: Prof. Dr. Takayuki Doi
Faculty of Pharmaceutical Sciences, Tohoku University
B.S.: April, 2005–March, 2009
Supervisor: Prof. Dr. Takayuki Doi

K. Ohsawa, *Total Synthesis of Thielocin B1 as a Protein–Protein Interaction Inhibitor of PAC3 Homodimer*, Springer Theses, DOI 10.1007/978-4-431-55447-9

The manufacturer's authorised representative in the EU is Springer Nature Customer Service Centre GmbH, Europaplatz 3, 69115 Heidelberg, Germany. If you have any concerns regarding our products, please contact ProductSafety@springernature.com

Printed and bound by CPI Group (UK) Ltd, Croydon, CR0 4YY
15/07/2026
02167654-0002